Joel Dorman Steele

The Story of the Stars

New descriptive astronomy

Joel Dorman Steele

The Story of the Stars
New descriptive astronomy

ISBN/EAN: 9783337410131

Printed in Europe, USA, Canada, Australia, Japan

Cover: Foto ©berggeist007 / pixelio.de

More available books at **www.hansebooks.com**

NEW

DESCRIPTIVE

ASTRONOMY,

BY

JOEL DORMAN STEELE, Ph.D.,

AUTHOR OF THE FOURTEEN-WEEKS SERIES IN NATURAL SCIENCE.

"The heavens declare the glory of God; and the firmament showeth is handiwork."—Psalm XIX, 1.

A POPULAR SERIES

IN

NATURAL SCIENCE,

BY

J. DORMAN STEELE, PH.D., F.G.S.,

Author of the Fourteen Weeks Series in Natural Science, etc., etc.

New Popular Chemistry.   New Descriptive Astronomy.

New Popular Physics.   New Hygienic Physiology.

New Popular Zoology.   Popular Geology.

An Introduction to Botany.

The Publishers can supply (to Teachers only) a Key containing Answers to the Questions and Problems in Steele's entire Series.

———————————

BARNES' HISTORICAL SERIES,

ON THE PLAN OF

STEELE'S FOURTEEN-WEEKS IN THE SCIENCES.

A Brief History of the United States.

A Brief History of France.

A Brief History of Ancient Peoples.

A Brief History of Mediæval and Modern Peoples

A Brief General History.

A Brief History or Greece.

A Brief History of Rome.

A Popular History of the United States.

PREFACE TO THE FIRST EDITION.

DURING the past few years great advances have been effected in astronomical science. Physics has come to the help of Mathematics, and, not content with the old question, where the heavenly bodies are, has sought to find out what they are. Valuable discoveries have been made concerning Meteors, Shooting Stars, the Constitution of the Sun, the Motion of the Heavenly Bodies, &c. The investigations connected with Spectrum Analysis have been especially suggestive. On every hand the facts of the New Astronomy have been accumulating. Until recently, however, they were scattered through many expensive books, and were consequently beyond the reach of the most of our schools. It has been the aim to collect in this little volume the most interesting features of the larger works.

Believing that Natural Science is full of fascination, the author has sought to weave the story of those far-distant worlds into a form that may attract the attention and kindle the enthusiasm of the pupil.

This work is not written for the information of scientific men, but for the inspiration of youth.

Therefore the pages are not burdened with a multitude of figures which no memory could retain.

Mathematical tables and data, Questions for Review, a very valuable Guide to the Constellations, and an Apparatus for Illustrating Precession, are given in the Appendix, where they may be useful for reference.

Those persons having a small telescope will find valuable assistance in the " List of interesting Objects for a common Telescope." The Index contains the pronunciation of many difficult names.

Particular attention is called to the method of classifying the measurements of Space, and the practical treatment of the subjects of Parallax, Harvest Moon, Eclipses, the Seasons, Phases of the Moon, Time, Nebular Hypothesis, Spectrum Analysis, and Precession.

To teachers hitherto compelled to use a cumbersome set of charts, it is hoped that the star maps here offered will present a welcome substitute. The geometrical figures, showing the actual appearance of the constellations, will relieve the mind confused with the idea of numberless rivers, serpents, and classical heroes. Only the brightest stars are given, since in practice it is found that pupils remember the general outlines alone, while the details are soon forgotten.

Many of the cuts are copied from the French edition of Guillemin's " Heavens." Acknowledgment for much valuable material is hereby made to this excellent work, and also to "Chambers's Astronomy," "Newcomb's Astronomy," and Young's "The Sun."

Finally, the author commits this little work to the hands of the young, to whose instruction he has consecrated the energies of his life, in the earnest hope that, loving Nature in all her varied phases, they may come to understand somewhat of the wisdom, power, beneficence, and grandeur displayed in the Divine harmony of the Universe.

> "One God, one law, one element,
> And one far-off Divine event
> To which the whole creation moves."

READING REFERENCES.

Chambers's Astronomy.—Young's The Sun.—Ball's Elements of Astronomy.—Newcomb's Popular Astronomy.—Lockyer's Spectrum Analysis.—Proctor's Other Worlds than Ours, Saturn, The Moon, Poetry of Astronomy, &c.—Delaunay's Cours D'Astronomie.—Haughton's Manual of Astronomy.—Newcomb and Holden's Astronomy.—Lockyer's Elements of Astronomy.—Norton's Spherical and Physical Astronomy. — Herschel's Outlines of Astronomy.—Robinson's Astronomy.—Mitchell's Popular Astronomy. — Arago's Popular Astronomy. — Airy's Lectures on Astronomy.—Hind's Solar System, and Introduction to Astronomy.—Lockyer's Elementary Lessons in Astronomy.—Proctor's Star Atlas.—Heis's Star Atlas.—Peck's Popular Astronomy.—Gillet and Rolfe's Astronomy.—Sharpless and Phillips's Astronomy.—Peabody's Elements of Astronomy.—Schellen's Spectrum Analysis.—Winchell's World-Life (excellent reading in connection with the Nebular Hypothesis).—Flammarion's Wonders of the Heavens.—Guillemin's The Heavens, revised by Proctor.—Loomis's Elements of Astronomy.—Proctor's Easy Star Lessons.—Olmstead's Letters on Astronomy.—Routledge's History of Science.—Buckley's History of Natural Science.—Williamson's Problems on the Globes.—The Popular Science Monthly (1872–1884).—Rambosson's Histoire Des Astres.

SUGGESTIONS TO TEACHERS.

THIS work is designed to be recited in the topical method. On hearing the title of a paragraph, the pupil should be able to draw upon the blackboard the diagram, and to state the substance of what is contained in the book. It will be noticed that *the order of topics*, in treating of the planets and also of the constellations, is uniform. If, each day, a portion of the class write their topics in full upon the blackboard, it will be found a valuable exercise in spelling, punctuation, and composition. Every point which can be illustrated in the heavens should be shown to the class. No description or apparatus can equal the reality in the sky. After a constellation has been traced, the pupil should be practised in star-map drawing.

The article on "Celestial Measurements," near the close of the work, should be constantly referred to during the term. In the figures, and especially in the star-maps, it should be remembered that the right-hand side represents the west ; and the left-hand, the east. To obtain this idea correctly, the book should, in general, be held up toward the southern sky.

For the purpose of more easily finding the heavenly bodies at any time, Whitall's Movable Planisphere is of great service. It may be procured of the publishers of this work. A tellurian is invaluable in explaining Precession of the Equinoxes, Eclipses, Phases of the Moon, etc. Messrs. A. S. Barnes & Co., New York City, furnish a good instrument at a low price. A small telescope, or even an opera-glass, will be useful. A good star-map, and as many advanced works upon Astronomy as can be secured, should be included in the teacher's outfit.

The pupil should, at the outset, get a distinct idea of the circles and planes of the celestial sphere. The subject of angular measurements can easily be made clear in this relation. A circle contains 360°; 90° reach from horizon to zenith; 180° produce opposition; while smaller distances can be shown in the sky (see pp. 216, 228).

Never let a pupil recite a lesson, nor answer a question, except it be a mere definition, *in the language of the book.* The text is designed to interest and instruct the pupil; the recitation should afford him an opportunity of expressing what he has learned, in his own style and words.

Teachers desiring additional information are advised to read "Newcomb's Astronomy," Young's "The Sun," Proctor's Works, "Chambers's Astronomy," and Ball's " Elements of Astronomy."

TABLE OF CONTENTS.

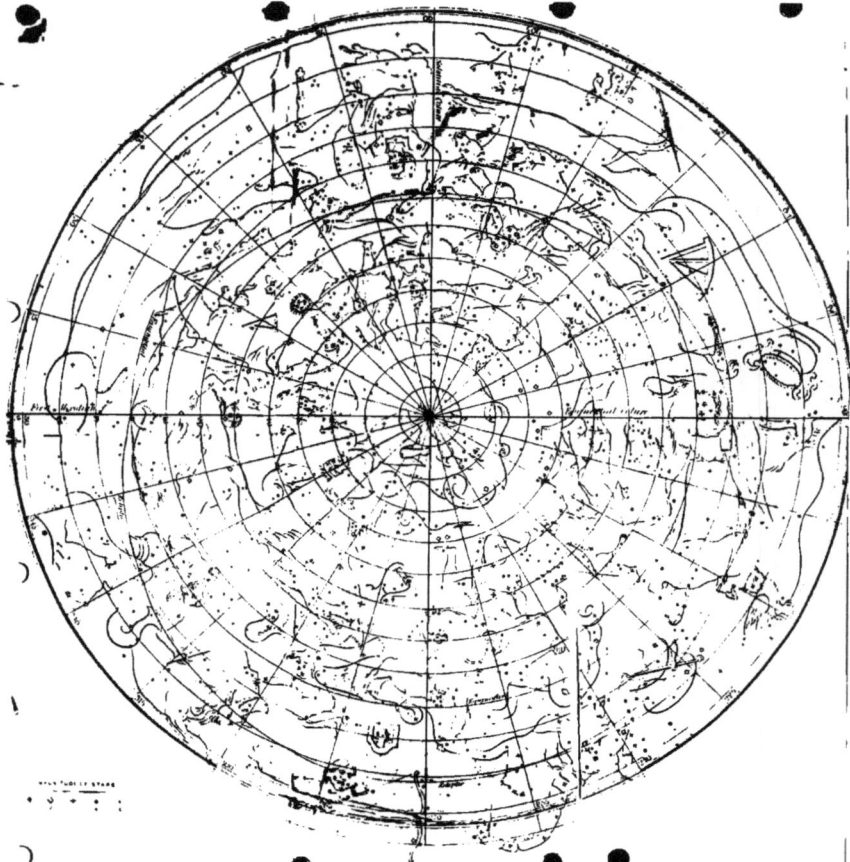

INTRODUCTORY REMARKS.*

ASTRONOMY (*astron*, a star ; *nomos*, a law) treats of the Heavenly Bodies—the sun, moon, planets, stars, etc., and, as our globe is a planet, of the earth also. It is, above all others, a science that cultivates the imagination. Yet its theories and distances are based upon rigorous mathematical demonstrations. Thus the study has at once the beauty of poetry and the exactness of Geometry.

The great dome of the sky, filled with glittering stars, is one of the most sublime spectacles in nature. To enjoy this fully, a night must be chosen when the air is clear, and the moon is absent. We then gaze upon a deep blue, an immense expanse studded with stars of varied color and brilliancy. Some shine with a vivid light, perpetually changing and twinkling ; others, more constant, beam tranquilly and softly upon us ; while many just tremble into our sight, like a wave that, struggling to reach some far-off land, dies as it touches the shore.

In the presence of such weird and wondrous beauty, the tenderest sentiments of the heart are aroused. A feeling of awe and reverence, of softened melancholy mingled with a thought of God, comes over us, and awakens the better nature within us. Those far-off lights seem full of meaning to us, could we but read their message ; they become real and sentient, and, like the soft eyes in pictures, look lovingly and inquiringly upon us. We come into communion with another life, and the soul asserts its immortality more strongly than ever before. We are humbled as we gaze upon the infinity of suns, and strive to comprehend

* This Introduction is designed merely to furnish suggestive material for conversation at the first lesson, preparatory to beginning the study. It is not intended for committal. Other topics may be found in the Questions given in the Appendix.

their enormous distances, and their magnificent retinue of worlds. The powers of the mind are aroused, and eager questionings crowd upon us. What are those glittering fires? What is their distance? Are they worlds like our own? Do living, thinking beings dwell upon them? Are they promiscuously scattered through space, or is there a system in the universe? Can we ever hope to fathom those mysterious depths, or are they closed to us forever?

Some of these problems have been solved; others yet await the astronomer whose eye shall be keen enough to read the mysterious scroll of the heavens. Two hundred generations of study have revealed to us such startling facts, that we wonder how man in his feebleness can grasp so much, see so far, and penetrate so deeply into the mysteries of the universe. Astronomy has measured the distance of a few stars, and of all the planets; computed the mass, size, days, years, seasons, and many physical features of the planets; made a map of the moon; tracked many of the comets in their immense sidereal journeys; and, at last, analyzed the structure of the sun and stars, and announced the very elements of which they are composed.

Observing for several evenings those stars which shine with a clear, steady light, we notice that they change their position with respect to the others. They are therefore called *planets* (literally *wanderers*). Others remain immovable, and shine with a shifting, twinkling light. They are termed the *fixed stars*, although it is now known that they also are in motion—the most rapid of any known even to Astronomy—but through such immense orbits that they seem to us to be stationary. Then, too, diagonally girdling the heavens, is a whitish, vapory belt—the *Milky Way*. This is composed of multitudes of millions of suns—of which our own sun itself is one—so far removed from us that their light mingles, and makes only a fleecy whiteness.

This magnificent panorama of the heavens is before us, inviting our study, and waiting to make known to us the grandest revelations of science.

I.

INTRODUCTION.

INTRODUCTION.

1. HISTORY.

1. AMONG THE CHINESE.

2. AMONG THE CHALDEANS.

3. AMONG THE GRECIANS.
   1. Thales.
   2. Anaximander.
   3. Pythagoras.
   4. Anaxagoras & Eudoxus.
   5. Hipparchus.

4. THE EGYPTIANS.
   1. The School at Alexandria.
   2. Ptolemy and his Theory.

5. THE SARACENS.

6. ASTROLOGY.

7. THE COPERNICAN SYSTEM.

8. TYCHO BRAHE.

9. KEPLER'S LAWS.

10. GALILEO.
    1. His Telescope.
    2. His Discoveries.
    3. Their Reception.

11. NEWTON, AND THE LAW OF GRAVITATION.
    a. Laws of Motion.
    b. Their Application to Moon's Phases.
    c. The Result.

2. SPACE.

1. CELESTIAL SPHERE.

2. THE THREE SYSTEMS OF CIRCLES.
   1. The Horizon.
      a. The Principal Circle.
      b. The Subord. Circle.
      c. Points.
      d. Measurements.
   2. The Equinoctial.
      a. The Principal Circle.
      b. The Subord. Circle.
      c. Points.
      d. Measurements.
   3. The Ecliptic.
      a. The Principal Circle.
      b. The Subord. Circle.
      c. Points.
      d. Measurements.

3. THE ZODIAC.

Fig. 2.

Sir Isaac Newton.

I.—THE HISTORY.

Astronomy is the most ancient of the sciences. The study of the stars is doubtless as old as man himself, and hence many of its discoveries date back of authentic records, amid the mysteries of tradition. In tracing its history, we shall speak only of those prominent facts that will enable us to understand its progress and glorious achievements.

The Chinese boast much of their astronomical discoveries. Indeed, their emperor claims a celestial ancestry, and styles himself the Son of the Sun. They possess an account of a conjunction of four planets and the moon, which occurred in the 25th century before Christ. They have also the first record of an eclipse of the sun (B.C. 2128); and one of their emperors put to death the chief astronomers Ho and Hi for failing to announce the solar eclipse of 2169 B.C.

The Chaldeans.—The Chaldean shepherds, watching their flocks by night under a sky famed for its clearness and brilliancy, could not fail to become familiar with many of the movements of the heavenly bodies. Their priests were astronomers; and their temples, observatories. When Alexander took Babylon (B.C. 331), he found a record of their observations reaching back nineteen centuries.* The Chaldeans divided the day into hours, invented the sun-dial, and discovered the Saros, or Chaldean Period—the length of time in which eclipses of the sun and the moon repeat themselves in the same order.

The Grecians.—Though the Asiatics were patient observers, they did not classify their knowledge, and lay the basis of a science. This became the work of the western mind.

THALES (B.C. 640–548), one of the seven sages of

* Many astronomical inscriptions have been found in the ruins of Nineveh. In the public library of that city there was a series of about seventy-two volumes, called the Observations of Bel. One book treated of the polar star (then Alpha of the Dragon), another of Venus, and a third of Mars. The earliest of these records are thought to date back as far as 2540 B.C. (See Records of the Past, Vol. 1.)

Greece, has been styled the Father of Astronomy. He taught that the earth is round, and that the moon receives her light from the sun. He determined when the equinoxes and the solstices occur, and also predicted an eclipse of the sun that is famous for having terminated a war between the Medes and the Lydians. These nations were engaged in a fierce battle, but the awe produced by the darkening of the sun was so great, that both sides threw down their arms and made peace.

ANAXIMANDER (B.C. 610–546) invented the sun-dial, and explained the cause of the moon's phases.

PYTHAGORAS (B.C. 570–500) founded a celebrated astronomical school at Crotona, Italy, where were educated hundreds of enthusiastic pupils.* He was emphatically a dreamer. He conceived a system of the universe, in many respects correct; yet he advanced no proof, made few converts to his views, and they were soon well-nigh forgotten.

He held that the sun is the center of the solar system, the planets revolving about it in circular orbits; that the earth rotates daily on its axis, and revolves yearly round the sun; that Venus is both morning and evening star; that the planets are placed at intervals corresponding to the scale in music, and that they move in harmony, making the "music of the spheres," but that this celestial concert is heard only by the gods,—the ears of man being too gross for such divine melody. He also believed that the planets are inhabited, and he even attempted to calculate the size of the animals in the moon.

* See Barnes's History of the Ancient Peoples, p. 174.

ANAXAGORAS (B.C. 500–428) taught that there is but one God, and that the sun is only a fiery globe, and should not be worshipped. He attempted to explain eclipses and other celestial phenomena by natural causes, saying that there is no such thing as chance or accident, these being only names for unknown laws. For his audacity and impiety, as his countrymen considered it, he and his family were doomed to perpetual banishment.

EUDOXUS, who lived in the fourth century B.C., invented the theory of the Crystalline Spheres. He held that the heavenly bodies are set, like gems, in hollow, transparent, crystal globes, which are so pure that they do not obstruct our view, while they all revolve around the earth ; and that the planets are placed in one globe, but have a power of moving themselves, under the guidance—as Aristotle taught —of a tutelary genius, who resides in each, and rules over it as the mind rules over the body.

HIPPARCHUS, who flourished in the second century B.C., has been called the Newton of Antiquity. He was the most celebrated of the Greek astronomers. He calculated the length of the year to within six minutes, discovered the precession of the equinoxes, and made the first catalogue of the stars—1080 in number.

The Egyptians.—Egypt, as well as Chaldea, was noted for its knowledge of the sciences long before they were cultivated in Greece. It was the practice of the Greek philosophers, before aspiring to the rank of teacher, to travel for years through these countries, and gather wisdom at its fountain-head. Pythagoras spent thirty years in this kind of study.

Two hundred years after Pythagoras, the celebrated school of Alexandria was established.* Here
were concentrated in vast libraries and princely
halls nearly all the wisdom and learning of the
world. Here flourished the sciences and arts, under
the patronage of munificent kings.

At this school, Ptolemy (A.D. 70), a Grecian, wrote
his great work, the Almagest, which for fourteen
centuries was the text-book of astronomers. In this
work was given what is known as the Ptolemaic
System. It was founded largely upon the materials
gathered by previous astronomers, such as Hipparchus, whom we have already mentioned, and Eratosthenes, who computed the size of the earth by the
means even now considered the best—the measurement of an arc of the meridian.

Ptolemaic System.—To the early astronomers, the
movements of the planets seemed extremely com-
plex. Venus, for instance, was sometimes seen as
evening star in the west, and then again as morning
star in the east. Sometimes she appeared to be
moving in the same direction as the sun, then, going
apparently behind the sun, she seemed to pass on
again in a course directly opposite. At one time,
she would recede from the sun more and more slowly
and coyly, until she would appear to be entirely stationary ; then she would retrace her steps, and seem
to meet the sun.

An attempt was made to account for all these
facts by an incongruous system of " Cycles and

* See Barnes's General History, p. 154.

epicycles," as it is called.* The advocates of this
theory assumed that every planet revolves in a
circle, and that the earth is the fixed center around
which the sun and the heavenly bodies move. They
then conceived that a bar, or something equivalent,
is connected at one end with the earth ; that at some
part of this bar the sun is attached ; while between
that and the earth, Venus is fastened—not to the bar
directly, but to a sort of crank ; and further on, Mer
cury is hitched on in the same way.

In Fig. 3, let A be the earth ; S, the sun ; A B D F,
the bar (real or imaginary) ; B C, the short bar or
crank to which Venus is tied ; D E, another bar for
Mercury ; F G, a fourth bar, with still another short
crank, at the end of which, H, Mars is attached.

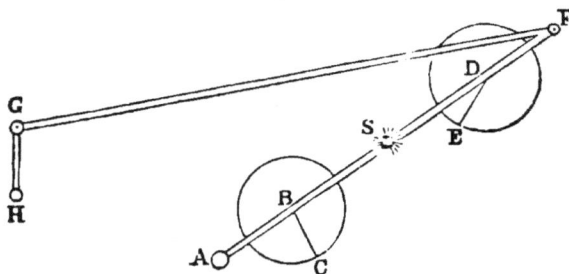

Fig. 3.

The Ptolemaic System.

Thus they had a complete system. They did not
exactly understand the nature of these bars—
whether they were real or only imaginary—but they
did comprehend their action, as they thought ; and

* Milton refers to this when he speaks of the heavens as—

" With centric and eccentric scribbled o'er,
Cycle and epicycle, orb in orb."

so they supposed the bar revolved, carrying the sun and planets along in a large circle about the earth ; while all the short cranks kept flying around, thus sweeping each planet through a smaller circle.

By this theory, we can see that the planets would sometimes go in front of the sun and sometimes behind ; and their places were so accurately predicted, that the error could not be detected by the rude instruments then in use. As soon as a new motion of one of the heavenly bodies was discovered, a new crank, and of course a new circle, was added to account for the fact. Thus the system became more and more complicated, until, at last, a combination of five cranks and circles was necessary to make the planet Mars keep pace with the Ptolemaic theory. No wonder that Alfonso, of Castile, a celebrated patron of Astronomy, revolted at the cumbersome machinery, and cried out, " If I had been consulted at the Creation, I could have done the thing better than that."

The Saracens. — After the destruction of the library at Alexandria, learning found a home among the Mohammedans. Bagdad on the Tigris, and Cordova on the Guadalquiver became centers of science, literature, and art. The treasures of Grecian knowledge were eagerly gathered by the Caliphs, and we are told that it was not uncommon to see, entering the gates of Bagdad, a whole train of camels loaded with Greek manuscripts. Gerbert, afterward Pope Sylvester II., learned the elements of astronomy at the University of Cordova, going, after the custom of the time, to Spain for instruc-

tion, as, formerly, philosophers had gone to Egypt
In the Moorish schools, geography was already
taught by the use of the globe.   The first observatory in Europe was erected at
Seville (1196).   The fragments
of Saracenic learning that have
come down to us show that
the Arabs had constructed astronomical tables, and endeavored
to perfect them by means of systematic observation of the
heavens. With the down-
fall of the Moors, and
the Revival of Learning,
Spain ceased to take the
lead in scientific study.

Fig. 4.

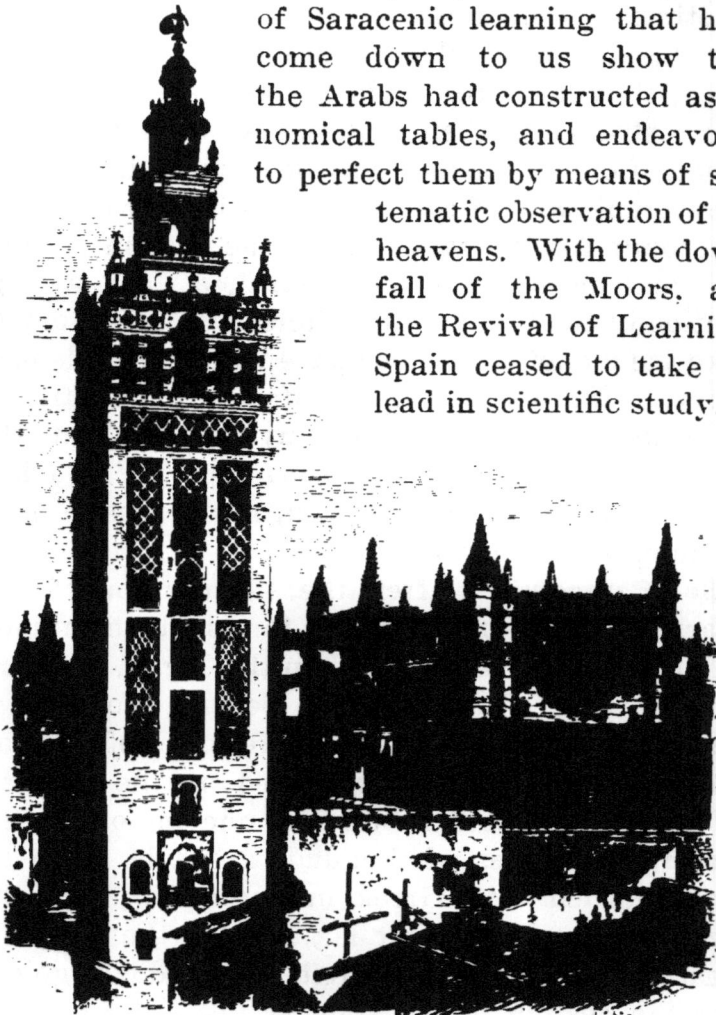

The Giralda, Moorish Observatory at Seville.

Astrology.—During all these centuries, astronomy owed its development quite as much to a desire of foretelling the future, as to a love for science. It was the prevalent belief that the stars rule the destinies of men. The Chaldeans scanned the heavens for purposes of divination, so that Chaldean and astrologer became synonymous. Tiberius, Emperor of Rome, practised astrology. Hippocrates himself, the Father of Medicine (B.C. 470), ranked this among the most important branches of knowledge for the physician. The mysterious study possessed a peculiar fascination for the Arabians, and they cultivated it assiduously. The Moorish astronomers were astrologers as well, and popularized the art in western Europe. This superstition reached the height of its influence during the Middle Ages.

The issue of any important undertaking and the fortunes of an individual were foretold by the astrologer, who drew up a *Horoscope* representing the position of the sun, moon, and planets at the beginning of the enterprise, or at the birth of the person. It was a complete and complicated system, and contained regular rules, which guided the interpretation, and which were so abstruse as to require years for their mastery. Venus foretold love; Mars, war; the Pleiades (Plē'-ya-dēz), storms at sea.

The ignorant were not the only dupes of this visionary system. Lord Bacon believed in it most firmly. Kepler, by casting nativities, eked out his miserable pittance as royal astronomer. So late even as the reign of Charles II., Lilly, a famous astrologer, was called before a committee of the House of

Commons, to give his opinion on the probable issue of some enterprise then under consideration.

However foolish the system of Astrology may have been, it preserved the science of Astronomy during the Dark Ages, and prompted to accurate observation and diligent study of the heavens.

The Copernican System.—About the commencement of the sixteenth century, Copernicus, breaking away from the theory of Ptolemy, that was still taught in the institutions of learning in Europe, revived the theory of Pythagoras. He saw how beautifully simple is the idea of considering the sun the grand center about which revolve the earth and the planets. He noticed how constantly, when we are riding swiftly, we forget our own motion, and think that the trees and fences are gliding by us in the contrary direction. He applied this thought to the movements of the heavenly bodies, and maintained that, instead of all the starry host revolving about the earth once in twenty-four hours, the earth simply turns on its own axis, and thus produces the apparent daily revolution of the sun and stars ; while the yearly motion of the earth about the sun, transferred in the same manner, would account for the solar movements.

Though Copernicus thus simplified the Ptolemaic theory, he yet found that the idea of circular orbits for the planets would not explain all the phenomena, and therefore retained the "cycles and epicycles" Alfonso had so heartily condemned. For forty years, this illustrious astronomer carried on his observations in the upper part of a humble, dilapidated farm-house, through the roof of which he had an unobstructed view of the sky. The work containing his theory was published just in time to be laid upon his death-bed.

Tycho Brahe, a celebrated Danish astronomer, next propounded a modification of the Copernican system. He rejected the idea of cycles and epicycles, but, influenced by certain passages of Scripture, maintained, with Ptolemy, that the earth is the center, and that all the heavenly bodies daily revolve about it in circular orbits. Brahe was a nobleman of wealth, and, in addition, received large sums of money from the government. He erected a magnificent observatory, and made many beautiful and rare instruments. Clad in his robes of state, he watched the heavens with the intelligence of a philosopher and the splendor of a king. His indefatigable industry and zeal resulted in the accumulation of a vast fund of astronomical knowledge, which, however, he lacked the ability to apply to any further advance in science.

His pupil, Kepler, saw these facts, and in his fruitful mind they germinated into three great truths, called Kepler's laws. These form one of the most precious conquests of the human mind. They are the three arches of the bridge over which Astronomy crossed the gulf between the Ptolemaic and Copernican systems.

Kepler's Laws.—Kepler, taking the investigations of his master, Tycho Brahe, determined to find what is the exact shape of the orbits of the planets. He adopted the Copernican theory—that the sun is the

center of the system.   At that time, all believed the
orbits to be circular.   They reasoned thus : the circle
is perfect ; it is the most beautiful figure in nature ;
it has neither beginning nor ending ; therefore, it is
the only form worthy of God, and He must have
used it for the orbits of the worlds He has made.

Imbued with this romantic view, Kepler commenced with a rigorous comparison of the places of
the planet Mars as observed by Brahe, with the
places as stated by the best tables that could be computed on the circular theory.   For a time, they
agreed, but in certain portions of the orbit the observations of Brahe would not fit the computed place
by eight minutes of a degree.   Believing that so
good an astronomer could not be mistaken as to the
facts, Kepler exclaimed, "Out of these eight minutes
we will construct a new theory that will explain the
movements of all planets."

He resumed his work, and for eight years continued to imagine every conceivable hypothesis, and
then patiently to test it—"hunt it down," as he
called it.   Each in turn proved false, until nineteen
had been tried.   He then determined to abandon the
circle and to adopt another form.   The *ellipse* suggested itself to his mind.   Let us see how this figure
is made.

Attach a thread to two pins, as at FF in the figure ;
next, move a pencil along with the thread, the latter
being kept tightly stretched, and the point will mark
a curve, flattened in proportion to the length of the
string,—the longer the string, the nearer a circle
will the figure become.   This figure is the *ellipse*.

The two points F F are called the *foci* (singular, *focus*). We can now understand Kepler's attempt, and the triumph which crowned his seventeen years of unflagging toil.

Fig. 5.

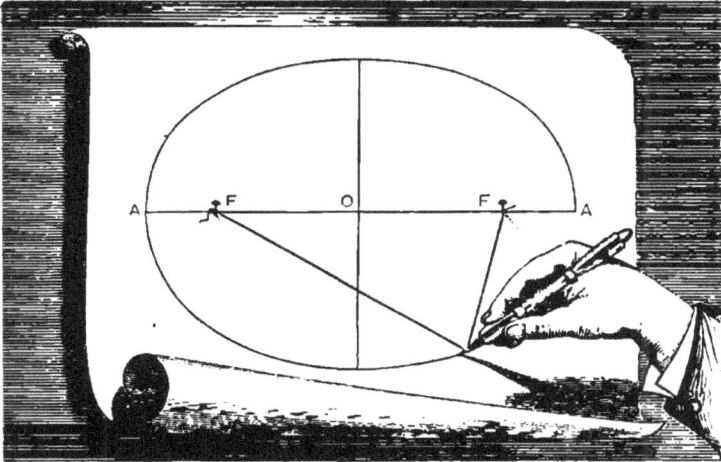

FIRST LAW.—With this figure he constructed an orbit having the sun at the center, and again followed the planet Mars in its course. But very soon there was as great a discrepancy between the observed and computed places as before. Undismayed by this failure, Kepler assumed another hypothesis, and determined to place the sun at one of the foci of the ellipse. Once more he "hunted down" the theory. For a whole year he traced the planet along the imaginary orbit, and it did not diverge. The truth was discovered at last, and Kepler (1609) announced his first great law—

Planets revolve in ellipses, with the sun at one focus.

SECOND LAW.—Kepler knew that the planets do
not move with equal velocity in the different parts
of their orbits.   He next set about establishing some
law by which this speed could be determined, and
the place of the planet computed.   He drew an
ellipse, and once more marked the various positions
of the planet Mars.   He soon found that when at
its *perihelion* (point nearest the sun) its motion
is fastest, but when at its *aphelion* (point furthest
from the sun) its motion is slowest.   Again he
"hunted down" various hypotheses, until, at last,

Fig. 6.

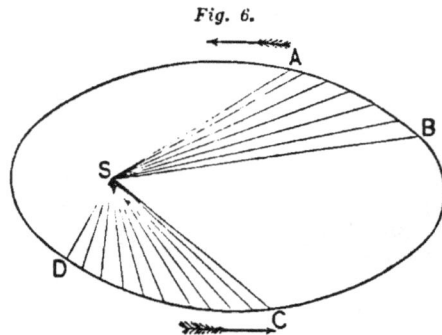

he discovered that though, in going from B to A, the
planet moves more slowly, and from D to C more
rapidly, yet the space inclosed between the lines SB
and SA is equal to that inclosed between SD and SC.
Hence the second law—

*A line connecting the center of the earth with the
center of the sun passes over equal spaces in equal
times.*

THIRD LAW.—Kepler, not satisfied with the discovery of these laws, now determined to ascertain if
there were not some relation existing between the

times of the revolutions of the planets about the sun and their distances from that body. With the same wonderful patience, he took the figures of Tycho Brahe, and began to compare them. He tried them in every imaginable relation. Next he took their squares, then he attempted their cubes. Here was the secret; but he toiled around it, made a blunder, and waited for months, until, once more, his patience triumphed, and he reached (1618) the third law—

*The squares of the times of revolution of the planets about the sun are proportional to the cubes of their mean distances from the sun.**

In rapture over the discovery of these three laws, so marked by that Divine simplicity which pervades all the laws of nature, Kepler exclaimed, "Nothing holds me. The die is cast. The book is written, to be read now or by posterity, I care not which. It may well wait a century for a reader, since God has waited six thousand years for an observer.†

Galileo.—Contemporary with Kepler was the great Florentine philosopher, Galileo. He discovered the laws of the pendulum and of falling bodies, as we have already learned in Physics. He was, however, educated in and believed the Ptolemaic system. A disciple of the Copernican theory happening to come to Pisa, where Galileo was teaching as professor in

---

* For example : The square of Jupiter's period is to the square of Mars's period, as the cube of Jupiter's distance is to the cube of Mars's distance ; or, representing the earth's time of revolution by P, and her distance from the sun by p, then letting D and d represent the same in another planet, we have the proportion $P^2 : D^2 : : p^2 : d^3$.

† Kepler, strangely enough, believed in the "Music of the Spheres." He made Saturn and Jupiter take the bass, Mars the tenor, Earth and Venus the counter, and Mercury the treble. This shows what a streak of folly or superstition may run through the character of the noblest man. However, as Johnson says, a mass of metal may be gold, though there be in it a little vein of tin.

the University, drew his attention to its simplicity
and beauty. His clear, discriminating mind perceived its perfection, and he henceforth advocated it
with all the ardor of his unconquerable zeal. Soon
after, he learned that one Jansen, a Dutch watchmaker, had invented a contrivance for making distant objects appear near. With his profound knowledge of optics and philosophical instruments, Galileo
caught the idea, and soon had a telescope completed.
It was a very simple affair—only a piece of lead pipe
with a lens set at each end; but it was destined to
overthrow the old Ptolemaic theory, and revolu-
tionize the science of Astronomy.

DISCOVERIES MADE WITH THE TELESCOPE.—Galileo
now examined the moon. He saw her mountains and
valleys, and watched the dense shadows upon her
plains. On January 8, 1610, he turned the telescope
toward Jupiter. Near it he saw three bright stars,
as he considered them, which were invisible to the
naked eye. The next night he noticed that they had
changed their relative positions. Astonished and
perplexed, he waited three days for a fair night in
which to resume his observations. The fourth night
was favorable, and he found the three stars had
again shifted. Night after night he watched them,
discovered a fourth star, and finally found that they
were rapidly revolving around Jupiter, each in its
elliptical orbit, with its own rate of motion, and all
accompanying the planet in its journey around the
sun. Here was a miniature Copernican system, hung
up in the sky for every one to see and examine for
himself.

RECEPTION OF THE DISCOVERIES.—Galileo met with the most bitter opposition. Many refused to look through the telescope lest they might become victims of the philosopher's magic. Some prated of the wickedness of digging out valleys in the fair face of the moon. Others doggedly clung to the theory they had held from their youth.* But the truth of the Copernican system was now fully established. Philosophers gradually adopted this view, and the Ptolemaic theory became a relic of the past.

Newton, a young man of twenty-four years, was spending the summer of 1666 in the country, on account of the plague which prevailed at Cambridge, his place of residence. One day, while sitting in a garden, an apple chanced to fall to the ground near him. Reflecting upon the strange power that causes all bodies thus to descend to the earth, and remembering that this force continues, even when we ascend to the tops of high mountains, the thought occurred to his mind, "May not this same force extend to a great distance out in space? Does it not reach the moon?"

LAWS OF MOTION.—To understand the reasoning that now occupied the mind of Newton, let us apply the laws of motion as we have learned them in

* As a specimen of the arguments adduced against the new system, the following by Sizzi is a fair instance. "There are seven windows in the head, through which the air is admitted to the body, to enlighten, to warm, and to nourish it,—two nostrils, two eyes, two ears, and one mouth. So in the heavens there are two favorable stars, Jupiter and Venus; two unpropitious, Mars and Saturn; two luminaries, the Sun and Moon; and Mercury alone, undecided and indifferent. From which, and from many other phenomena in Nature, such as the seven metals, etc., we gather that the number of planets is necessarily seven. Moreover, the satellites are invisible to the naked eye, can exercise no influence over the earth, and would be useless, and therefore do not exist. Besides, the week is divided into seven days, which are named from the seven planets. Now, if we increase the number of planets, this whole system falls to the ground."

Physics.   When a body is set in motion, it will continue to move forever in a straight line, unless another force is applied.   As there is no friction in space, the planets do not lose any of their original velocity, but move now with the same speed which they received at the beginning.   But this would make them all pass along straight lines, and not circular orbits.   What causes the curve ?   Obviously, another force.   For example : I throw a stone into the air.   It does not move in a straight line, but in a curve, because the earth constantly bends it downward.

APPLICATION.—Just so the moon is moving around the earth, not in a straight line, but in a curve.   Can it not be that the earth bends *it* downward, just as it does the stone ?   Newton knew that a stone falls toward the earth sixteen feet the first second.   He conceived, after a careful study of Kepler's laws, that the attraction of the earth diminishes according to the square of the distance.   He supposed (according to the measurement then received) that a body on the surface of the earth is exactly four thousand miles from the center.   He now applied this imaginary law.   Suppose the body is removed four thousand miles from the surface of the earth, or eight thousand miles from the center.   Then, as it is twice as far from the center, its weight will be diminished 2^2, or 4 times.   If it were placed 3, 4, 5, 10 times further away, its weight would then decrease 9, 16, 25, 100 times.   If, then, the stone at the surface of the earth (four thousand miles from the center) falls sixteen feet the first second, at eight thousand miles it would fall only four feet ; at 240,000 miles, or the distance of the moon, it would fall only about one-twentieth of an inch (exactly .053).

Next the question arose, "How far does the moon fall toward the earth, *i. e.*, bend from a straight line, every second ?" For sixteen years, with a patience rivaling Kepler's, this philosopher sought to solve the problem. He toiled over interminable columns of figures, to find how much the moon's path curves each second. At last, he reached a result, which was nearly, but not quite, exact. Disappointed, he laid aside his calculations. Repeatedly he reviewed them, but could not find a mistake. At length, while in London, he learned of a new and more accurate measurement of the distance from the circumference to the center of the earth. He hastened home, inserted this new value in his calculations, and soon found that the result would be correct. Overpowered by the thought of the grand truth just before him, his hand faltered, and he called upon a friend to complete the computation.

From the moon, Newton passed on to the other heavenly bodies, calculating and testing their orbits. Finally, he turned his attention to the sun, and, by reasoning equally conclusive, proved that the attraction of that great central orb compels all the planets to revolve about it in elliptical orbits, and holds them with an irresistible power in their appointed paths.*

---

* "Do not understand me at all as saying there is no mystery about the planets' motion. There is just one single mystery,—gravitation ; and it is a very profound one. How it is that an atom of matter can attract another atom, no matter how great the distance, no matter what intervening substance there may be ; how it will act upon it, or at least behave as if it acted upon it, —I do not know, I cannot tell. Whether they

At last, he announced this grand Law of Gravitation : *Every particle of matter in the universe attracts every other particle, of matter with a force directly proportional to its quantity of matter, and decreasing as the square of the distance increases.*

II. SPACE.

WE now in imagination pass into space, which stretches out in every direction, without bounds or measures. We look up to the heavens, and try to locate some object among the mazes of the stars. Bewildered, we feel the necessity of some system of measurement. Let us try to understand the one adopted by astronomers.

The Celestial Sphere.—The blue arch of the sky, as it appears to be spread over us, is termed the *Celestial Sphere.* There are two points to be noticed here.

First, that so far distant is this imaginary arch from us, that if any two parallel lines from different parts of the earth were drawn to this Sphere, they would apparently intersect. Of course, this could not be the fact ; but the distance is so immense, that we are unable to distinguish the little difference of

are pushed together by means of an intervening ether, or what is the action, I cannot understand. It stands with me along with the fact, that, when I will my arm to rise, it rises. It is inscrutable. All the explanations that have been given of it seem to me merely to darken counsel with words and no understanding. They do not remove the difficulty at all. If I were to say what I really believe, it would be, that the motion of the spheres of the material universe stand in some such relation to Him in whom all things exist, the ever-present and omnipotent God, as the motions of my body do to my will : I do not know how, and never expect to know."— *Prof. Young.*

four or even eight thousand miles, and the two lines would seem to unite : so we must consider this great earth as a mere speck or point at the center of the Celestial Sphere.

Second, that we must neglect the entire diameter of the earth's orbit, so that if we should draw two parallel lines, one from each end of the earth's orbit, to the Celestial Sphere, although these lines would be nearly 186,000,000 miles apart, yet they would appear to pierce the Sphere at the same point ; which is to say, that, at that enormous distance, 186,000,000 miles shrink to a point. Consequently, in all parts of the earth, and in every part of the earth's orbit, we see the fixed stars in the same place.

This sphere of stars surrounds the earth on every side. In the daytime, we cannot see the stars because of the superior light of the sun ; but, with a telescope, they can be traced, and an astronomer will find certain stars as well at noon as at midnight.

One half of the sphere is constantly visible to us ; and so far distant are the stars, that we see just as much of the sphere as we should if the upper part of the earth were removed, and we were to stand four thousand miles further away, or at the center of the earth, where our view would be bounded by a great circle of the earth.

On the concave surface of the Celestial Sphere, there are imagined to be drawn three systems of circles: the HORIZON, the EQUINOCTIAL, and the ECLIPTIC SYSTEM. Each of these has (1) its *Principal Circle*, (2) its *Subordinate Circles*, (3) its *Points*, and (4) its *Measurements*.

1. THE HORIZON SYSTEM.

(*a*) **The Principal Circle** is the *Rational Horizon.* This is the great circle whose plane, passing through the center of the earth, separates the visible from the invisible heavens. The *Sensible Horizon* is the small circle where the earth and the sky seem to meet: it is parallel to the rational horizon, but distant from it the semi-diameter of the earth. No two places have the same sensible horizon: any two, on opposite sides of the earth, have the same rational horizon.

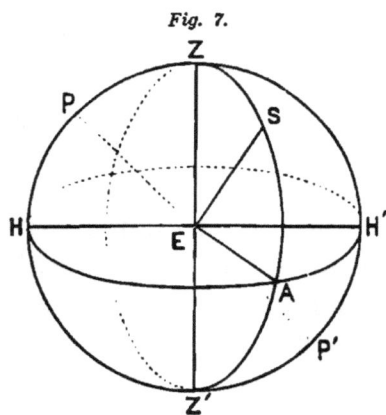

Fig. 7.

E, *center of earth*; Z, *zenith*; Z', *nadir*; PP', *axis of earth*; H AH', *horizon*; S, *a star*; Z S Z', *vertical circle passing through* S; AS, *altitude of star*; ZS, *zenith distance of star*; H'A, *azimuth of star*.

(*b*) **The Subordinate Circles** are the *Prime Vertical* circle, and the *Meridian.* A vertical circle is one passing through the poles of the horizon (zenith, and nadir). The Prime Vertical is a vertical circle passing through the East and West points The Meridian is a vertical circle passing through the North and South points.

(*c*) **The Points** are the *Zenith*, the *Nadir*, and the N., S., E., and W. points. The Zenith is the point directly overhead, and the Nadir, the one directly underfoot. They are also the poles of the horizon-- *i. e.*, the points where the axis of the horizon pierces the Celestial Sphere. The N., S., E., and W. points are familiar.

(*d*) **The Measurements** are *Azimuth, Amplitude, Altitude,* and *Zenith distance.*

AZIMUTH is the distance from the meridian, measured east or west, on the horizon, to a vertical circle passing through the object.

AMPLITUDE (the complement of Azimuth) is the distance from the Prime Vertical, measured on the horizon, north or south.

ALTITUDE is the distance from the horizon, measured on a vertical circle, toward the zenith.

ZENITH DISTANCE (the complement of Altitude) is the distance from the zenith, measured on a vertical circle, toward the horizon.

The Horizon system is one commonly used in observations with Mural Circles, and Transit Instruments.

2. THE EQUINOCTIAL SYSTEM.

(*a*) **The Principal Circle** is the *Equinoctial.* This is the *Celestial Equator,* or the earth's equator extended to the Celestial Sphere. At all places between the equator and the pole, the celestial equator is inclined to the horizon at an angle equal to the distance of the zenith of the place from the pole.*

(*b*) **The Subordinate Circles** are the *Hour Circles* (Right Ascension Meridians), the *Colures,* and the *Declination Parallels.*

---

* The latitude of a place is its distance from the equator, and this equals the distance of the zenith of the place from the equinoctial. Hence, having given the latitude of a place, to find the height of the celestial equator above its horizon, subtract the latitude from 90°, and the remainder is the required angular distance. In like manner, the latitude subtracted from 90° gives the co-latitude of the place -the complement of the latitude.

THE HOUR CIRCLES are thus located. The Equinoctial is divided into 360°, equal to twenty-four
hours of motion—thus making 15° equal to one hour
of motion. Through these divisions run twenty-four
meridians, each constituting an hour of motion
(time) or 15° of space. The Hour Circles may be
conceived as meridians of terrestrial longitude (15°
apart) extended to the Celestial Sphere.

THE COLURES are two principal meridians; the
Equinoctial Colure is the meridian passing through
the equinoxes; the *Solstitial Colure* is the meridian
passing through the solstitial points.

THE DECLINATION PARALLELS are small circles
parallel to the Equinoctial; or they may be conceived
as the parallels of terrestrial latitude extended to the
Celestial Sphere.

(c) **The Points** are the *Celestial Poles*, and the
Equinoxes.

THE CELESTIAL POLES are the points where the
axis of the earth extended pierces the Celestial
Sphere, and are the extremities of the celestial axis,
as the poles of the earth are the extremities of the
earth's axis.   The North Pole is marked very nearly
by the North Star, and every direction *from* that is
reckoned south, and every direction *toward* that is
reckoned north, however it may conflict with our
ideas of the points of the compass.

THE EQUINOXES are the points where the Equinoctial and the Ecliptic (the sun's apparent path through
the heavens) intersect.

(d) **The Measurements** are *Right Ascension* (R.A.),
Declination, and *Polar Distance*.

RIGHT ASCENSION is distance from the Vernal Equinox, measured on the equinoctial eastward to the meridian which passes through the body. R. A. corresponds to terrestrial longitude, and may extend to 360° East, instead of 180° as on the earth. R. A. is never measured westward. The starting point is the meridian passing through the vernal equinox, as the meridian passing through Greenwich is the point from which terrestrial longitude is measured.

DECLINATION is distance from the equinoctial, measured on any Hour Circle or meridian north or south. It corresponds to terrestrial latitude.

POLAR DISTANCE (the complement of Declination) is the distance from either Pole, measured on an Hour Circle.

The Equinoctial System is largely used by modern astronomers, and accompanies the Equatorial Telescope, Sidereal Clock, and Chronographs of the best Observatories.

3. THE ECLIPTIC SYSTEM.

(a) **The Principal Circle** is the *Ecliptic.* This is the apparent path of the sun in the heavens. It is inclined to the equinoctial $23\frac{1}{2}°$ ($23°\ 27'\ 15''$, Jan. 1, 1884), which measures the inclination of the Earth's Equator to its orbit, and is called the *obliquity of the ecliptic.* (See p. 58.)

The inclination of the ecliptic to the horizon, unlike that of the equinoctial, varies at different times of the year. The angle that the ecliptic makes with the horizon is greatest when the vernal equinox is on the western horizon and the autumnal on the eastern ; it is least when the vernal equinox is on the eastern horizon and the autumnal on the western.*

(*b*) **The Subordinate Circles** are *Circles of Celestial Longitude,* and *Parallels of Celestial Latitude.*

The Circles of Celestial Longitude are now seldom employed. They are measured on the Ecliptic, as circles of Right Ascension (R. A.) are measured on the Equinoctial.

The Parallels of Celestial Latitude are little used. They are small circles drawn parallel to the ecliptic, as parallels of declination are drawn parallel to the equinoctial.

(*c*) **The Points** are the *Poles of the Ecliptic,* the *Equinoxes,* and the *Solstices.*

The Poles of the Ecliptic are the points where the axis of the earth's orbit meets the Celestial Sphere.

The Equinoxes are the points where the ecliptic intersects the equinoctial. The place where the sun crosses the equinoctial † in going north, which occurs about the 21st of March, is called the Vernal Equinox. The place where the sun crosses the equinoctial in going south, which occurs about the 21st of September, is called the Autumnal Equinox. The *Solstices* are the two points of the ecliptic most distant from the Equator ; or they may be considered to mark the sun's furthest declination north and south of the equinoctial. The Summer Solstice occurs about the

* In the former instance, the angle is equal to the co-latitude, plus 23¼° (the inclination of the ecliptic to the equinoctial) ; and, in the latter, the co-latitude minus 23¼°. Thus, at the latitude of New York, it varies from 90° — 41° + 23¼° = 72¼° ; to 90° — 41° — 23¼° = 25¼°. In the one case, the summer solstice is on the meridian of the place, and, in the other, the winter.

† " This is commonly called ' crossing the line.' "

21st of June ; the Winter Solstice occurs about the 21st of December.

(*d*) **The Measurements** are *Celestial Longitude* and *Latitude.*

CELESTIAL LONGITUDE is distance from the Vernal Equinox measured on the ecliptic, eastward.

CELESTIAL LATITUDE is distance from the ecliptic measured on a Subordinate Circle, north or south.

THE ZODIAC.

A belt of the Celestial Sphere, 8° on each side of the ecliptic, is styled the *Zodiac.* This is of very high antiquity, having been in use among the ancient Hindoos and Egyptians. The Zodiac is divided into twelve equal parts—of 30° each—called Signs, to each of which a fanciful name is given. The following are the names of the

SIGNS OF THE ZODIAC.

Aries	♈	Libra	♎
Taurus	♉	Scorpio	♏
Gemini	♊	Sagittarius	♐
Cancer	♋	Capricornus	♑
Leo	♌	Aquarius	♒
Virgo	♍	Pisces	♓

" The first, ♈, indicates the horns of the Ram ; the second, ♉, the head and horns of the Bull ; the barb attached to a sort of letter, ♏, designates the Scorpion ; the arrow, ♐, sufficiently points to Sagittarius ; ♑ is formed from the Greek letters, τρ, the two first letters of τράγος, *a goat.* Finally, a balance, the flowing of water, and two fishes, tied by a string, may be imagined in ♎, ♒, and ♓, the signs of Libra, Aquarius, and Pisces." (See pp. 210, 295.)

PRACTICAL QUESTIONS.

1. How high is the North Star above your horizon ?

2. What is the sun's right ascension at the autumnal equinox ? At the vernal equinox ?

3. What was the first discovery made by the telescope ?

4. How high above the horizon of any place are the equinoctial points when they pass the meridian ?

5. Jupiter revolves around the sun in 12 of our years. Assuming the earth's distance from the sun to be 93,000,000 miles, compute Jupiter's distance by applying Kepler's third law.

6. The latitude of Albany is 42° 39′ N ; what is the sun's meridian altitude at that place when it is in the celestial equator ?

7. What is the co-latitude of a place ?

8. What is the declination of the zenith of the place in which you reside ?

9. Why are the stars generally invisible by day ?

10. Why is the ecliptic so called ?

11. Who first taught that the earth is round ?

12. What is Astrology ?

13. How can we distinguish the fixed stars from the planets !

14. How long was the Ptolemaic System accepted ?

15. In what respect did the Copernican System differ from the one now received ?

16. For what is Astronomy indebted to Galileo ?   To Newton ?

17. What is the amount of the obliquity of the ecliptic ?

18. Define Zenith. Nadir. Azimuth. Altitude. Equinoctial. Right Ascension. Declination. Equinox. Ecliptic. Colure. Solstice. Polar distance. Zenith distance. The Zodiac.

19. If the R. A. of the sun be 80°, state in what sign he is then located ? 160° ? 280° ?

20. Why does the angle which the ecliptic makes with the horizon vary ?

21. Why is the angle which the celestial equator makes with the horizon constant ?

II.

THE SOLAR SYSTEM.

———————

" In them hath He set a tabernacle for the sun."

" This world was once a fluid haze of light,
Till toward the center set the starry tides
And eddied into suns, that wheeling cast
The planets."—TENNYSON.

I. THE SUN

1. DISTANCE.
2. LIGHT & HEAT.
3. APPARENT SIZE.
4. REAL DIMENSIONS.
5. SOLAR SPOTS...
   - a. Discovery.
   - b. Number and Location.
   - c. Size.
   - d. Constituents.
   - e. Motion across Disk.
   - f. Change in Rate.
   - g. Prove the Rotation of Sun.
   - h. Synodic and Sidereal Rotation
   - i. Path of Spots.
   - j. Individual Motion.
   - k. Change in Form.
   - l. Periodicity of Spots.
   - m. Planetary Influence.
   - n. Influence on Terrestrial Heat, etc.
   - o. Heat of Spots.
   - p. Depression of Spots.
   - q. Brightness of Spots.
   - r. Faculæ, rice-grains, etc.
6. PHYSICAL CONSTITUTION....
   - a. Wilson's Theory.
   - b. Present Theory (Kirchhoff's).
7. HOW SOLAR HEAT IS PRODUCED.

II. THE PLANETS.

—INTRODUCTION...
   - a. Common Characteristics.
   - b. Comparison of Planets.
   - c. Properties of the Ellipse.
   - d. Planetary Orbits.
   - e. Comparative Size of Planets.
   - f. Conjunction of.
   - g. Are Planets Inhabited?
   - h to p. Division of Planets, etc.

1. VULCAN.
2. MERCURY......
   - a. Description.
   - b. Motion in Space.
   - c. Distance from Earth.
   - d. Dimensions.
   - e. Seasons.
   - f. Telescopic Features.
3. VENUS ....Repeat same Analysis as of Mercury.
4. THE EARTH ....
   - a. Dimensions.
   - b. Rotundity.
   - c. Apparent & Real Motion.
   - d. Diurn'l Motion of Earth.
     - 1. Diurnal Motion of Sun.
     - 2. Unequal rate of Motion.
     - 3. Orbits of Stars.
     - 4. Unequal Velocities of Stars.
     - 5. Appearance of Stars, etc.
   - e. Yearly Motion of Sun: its Consequences.
     - 1. Change in appearance of heavens.
     - 2. Yearly path of Sun.
     - 3. Moves N. & S.
     - 4. Change of Seasons, etc. 20 points under this topic.
   - f. Precession of Equinoxes.
   - g. Nutation.
   - h. Refraction & Aberration.
   - i. Parallax.

—THE MOON.
   - a. Motion.
   - b. Dimens'ns.
   - c. Librations.
   - d. L'g't & H't.
   - e. Cen.of Grav.
   - f. Atmosph're.
   - g. Lunarians.
   - h. Earth-shine.
   - i. Phases.
   - j. Harv'st M'n.
   - k. Wet Moon.
   - l. Nodes.
   - m. Occulat'n.
   - n. Seasons.
   - o. Telescopic Features.

—ECLIPSES.

—THE TIDES.

5. MARS........Same Analysis as Mercury.
6. THE MINOR PLANETS.
7. JUPITER......Same Analysis as Mercury.
8. SATURN.....   "   "
9. URANUS.....   "   ..
10. NEPTUNE.....   "   "

III. METEORS, AND SHOOTING STARS.)
IV. COMETS .... } The subjects of the paragraphs may be inserted
V. THE ZODIACAL LIGHT ....) by the pupil, to complete these analyses, at the pleasure of the teacher.

THE SOLAR SYSTEM.

INTRODUCTION.

THE **Solar System** is mainly comprised within the limits of the Zodiac. It consists of—

1. The Sun—the center.
2. The major planets—Vulcan (undetermined), Mercury, Venus, Earth, Mars, Jupiter, Saturn, Uranus, Neptune.
3. The minor planets, at present (1884) two hundred and thirty-seven in number.
4. The satellites, or moons, twenty in number, which revolve around the different planets.
5. Meteors and shooting-stars.
6. Thirteen comets, which have now been found, by a second return, to move, like the planets, in elliptic paths, and to revisit the sun periodically.
7. The Zodiacal Light.

How we are to imagine the solar system to ourselves.—We are to think of it as suspended in space; being held up, not by any visible object, but in accordance with the law of Universal Gravitation discovered by Newton, whereby each planet attracts every other planet and is in turn attracted by all.

First, the Sun, a great central globe, so vast as to overcome the attraction of all the planets, and compel them to circle around him; next, the planets, each turning on its axis while it flies around the sun in an elliptical orbit; then, accompanying these, the satellites, each revolving about its own planet, while all whirl in a dizzy waltz about the central orb; next, the comets, rushing across the planetary orbits at irregular intervals of time and space; and finally, shooting-stars and meteors darting hither and thither, interweaving all in apparently inextricable confusion.

To make the picture more wonderful still, every member is flying with an inconceivable velocity, and yet with such accuracy that the solar system is the most perfect timepiece known.

I. THE SUN.

Sign, ☉, a buckler with its boss.

Distance.—The sun's average distance from the earth is nearly 93,000,000 miles.* Since the earth's orbit is elliptical, and the sun is situated at one of its foci, the earth is 3,000,000 miles further from the sun in aphelion than in perihelion.

* The sun's distance from the earth is determined, as we shall learn hereafter (see Celestial Measurements), by means of the solar parallax. In the former editions of this work, the parallax of 8″.94—deduced principally from observations upon the planet Mars in 1862—was accepted. This gave a solar distance of about 91¼ million miles, and has been in general use among astronomers until recently. The observations of the last few years have, however, shown that the true parallax is smaller, and that the sun is a little further off than was supposed. Astronomers are not fully agreed upon the exact parallax that should be adopted, but there seems to be a general converging of opinion toward 8″.80 as being, if not the exact parallax, at least as near it as we are able at present to come. This new determination of the solar parallax renders necessary a corresponding change in the planetary distances, etc., as the sun's distance is the unit used by astronomers in making all celestial measurements. In this chapter, the author has followed the data given by Prof. Young in his work upon the Sun, as being the most recent and authoritative view. (See p. 280.)

As we attempt to locate the heavenly bodies in space, we are startled by the enormous figures employed. The first number, 93,000,000 miles, is far beyond our grasp. Let us, however, try to comprehend it.* If there were air to convey a sound from the sun to the earth, and a noise could be made loud enough to pass that distance, it would require over fourteen years for it to come to us. Suppose a railroad could be built to the sun. An express-train, traveling day and night, at the rate of thirty miles an hour, would require 352 years to reach its destination. Ten generations would be born and would die; the young men would become gray-haired, and their great-grandchildren would forget the story of the beginning of that wonderful journey, and would read it in history, as we now read of Queen Elizabeth or of Shakspere; the eleventh generation* would see the solar station at the end of the route. Yet this enormous distance of 93,000,000 miles is used as the unit for expressing celestial distances,—as the foot-rule for measuring space; and astronomers speak of so many times the sun's distance as we speak of so many feet or inches.

The Light of the Sun is equal to 5,563 wax-candles held at a distance of one foot from the eye. It would require 600,000 full-moons to produce a day as brilliant as one of cloudless sunshine.†

* If a babe were born with an arm long enough to reach the sun, and should touch that fiery globe, the infant would grow to manhood and to old age and finally die, before the sensation could traverse the nerve to his brain, and he feel the burn.

† According to Langley, the sun is blue, and to the inhabitants of other worlds may shine as a bluer star than Vega. The light from different parts of the solar disk, however, varies in color: while that from the center has a decidedly-blue tint, that from the edge is of a chocolate hue. This difference is probably owing to the fact that the latter

The Heat of the Sun.—The amount of heat we receive annually is sufficient to melt a layer of ice 110 feet thick, extending over the whole earth.* Yet the sunbeam is only $\frac{1}{300,000}$ part as intense as it is at the surface of the sun. Moreover, the heat and light stream off into space equally in every direction. Of this vast flood, only one twenty-three-hundred-millionth part reaches the earth.

If the heat of the sun were produced by the burning of coal, it would require a layer sixteen feet in thickness, extending over its whole surface, to feed the flame a single hour. Were the sun a solid body of coal, it would burn up at this rate in forty-six centuries. Sir John Herschel says that if a solid cylinder of ice 45 miles in diameter and 200,000 miles long were plunged, end first, into the sun, it would melt in a second of time.

Apparent Size.—The sun appears to be a little over half a degree in diameter, so that 337 solar disks, laid side by side, would make a half-circle of the celestial sphere. It seems a trifle larger to us in winter than in summer, as we are 3,000,000 miles nearer it. If we represent the luminous surface of the sun when at its average (mean) distance by 1,000, the same surface will be represented when in aphelion (July) by 967, and when in perihelion (January) by 1,034.

---

passes through a greater thickness of the solar atmosphere, while our own atmosphere does its part in strangling the blue rays of the sunlight, the red rays filtering through with little loss.

* Recent experiments by Langley seem to increase this estimate to that of a sheet of ice 180 feet thick (Popular Science Monthly, Sept., 1885).

Dimensions.—Its *diameter* is about 866,000 miles.*
Let us try to understand this amount by comparison.

A mountain upon the surface of the sun, to bear
the same proportion to the globe itself as the loftiest
peak of the Himalayas does to the earth, would need
to be about 600 miles high.

Again : Suppose the sun were hollow, and the
earth, as in the cut (Fig. 8), placed at the center, not

Fig. 8.

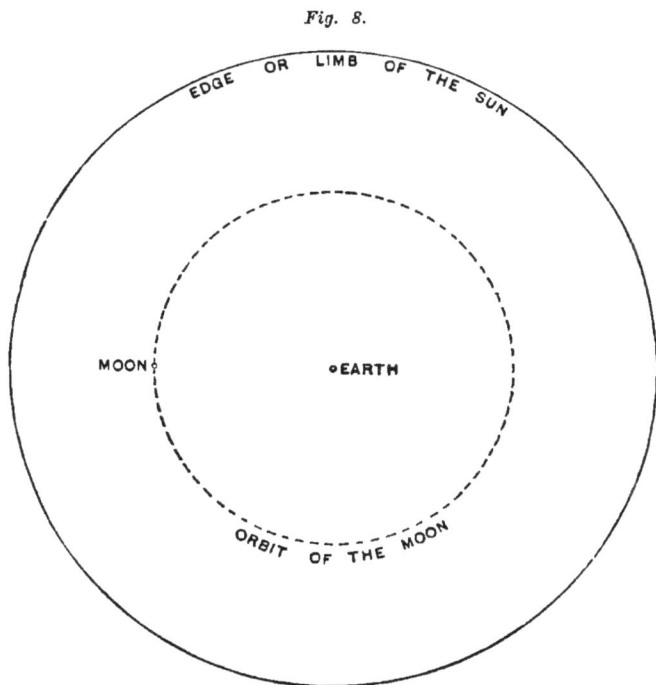

only would there be room for the moon to revolve in
its regular orbit within the shell, but that would

* Pythagoras, whose theory of the universe was in so many respects very like the
one we receive, believed the sun to be 44,000 miles from the earth, and seventy-five
miles in diameter.

stretch off in every direction nearly 200,000 miles beyond.

Its *volume* is 1,300,000 times that of the earth—*i. e.*, it would take 1,300,000 earths to make a globe the size of the sun. Its *mass* is 750 times that of all the planets and moons in the solar system, and 330,000 times that of the earth. Its *weight* may be expressed in tons, thus :

$$1,910,278,070,000,000,000,000,000,000.*$$

The *Density* of the sun is only about one-fourth that of the earth, or 1.41 that of water, so that the weight of a body transferred from the earth to the sun would not be increased in proportion to the comparative size of the two. On account also of the vast size of the sun, its surface is so far from its center that the attraction is largely diminished, since that decreases, we remember, as the square of the distance. However, a man weighing at the earth's equator 150 lbs., at the sun's equator would weigh about two tons,—a force of attraction that would instantly crush him. At the earth's equator, a stone falls 16 feet the first second ; at the sun's equator, it would fall 444 feet.†

Telescopic Appearance of the Sun : Sun Spots.—We may sometimes examine the sun at early morning or late in the afternoon with the naked eye, and

---

* This number is meaningless to our imagination, but yet it represents a force of attraction that holds our own earth and all the planets steadily in their places; while it fills the mind with an indescribable awe as we think of that Being who "made the sun, and holds it in the very palm of His hand."

† A singular consequence of this has been suggested. "A cannon-ball could be thrown only a short distance, since it would pass through a path of great curvature, and would fall to the sun within a few yards of the gun."

at midday by using a smoked glass. The disk will
appear distinct and circular, and with no spot to dim
its brightness. If we use a telescope of moderate

Fig. 9.

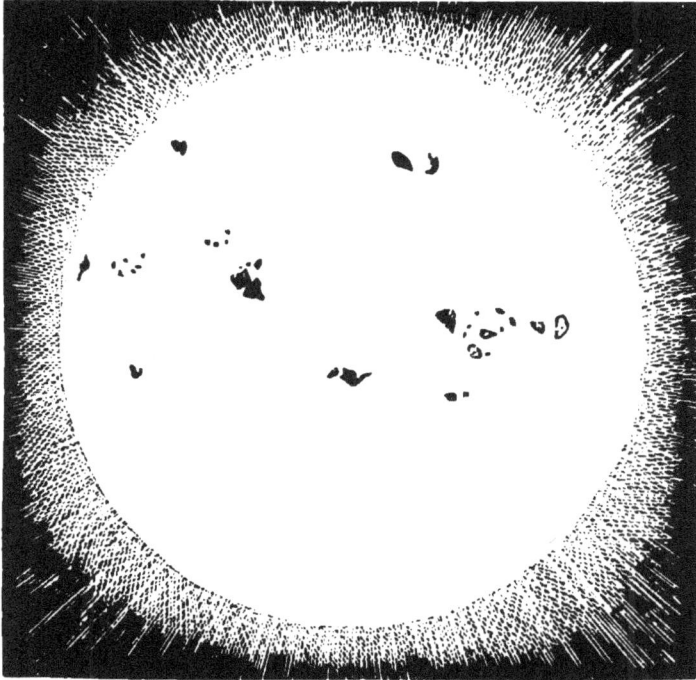

The Sun seen through a Telescope.

power, taking the precaution to shield the eye with
a colored eye-piece, we shall find the surface of the
sun sprinkled with irregular spots (Fig. 9).*

* The natural purity of the sun seems to have been formerly an article of faith among
astronomers, and therefore on no account to be called in question. Scheiner, it is said,
having reported to his superior that he had seen spots on the sun's face, was abruptly
dismissed with these remarks : " I have read Aristotle's writings from end to end many
times, and I assure you I do not find anything in them similar to that which you mention. Go, my son, tranquillize yourself ; be assured that what you take for spots are the
faults of your glasses or your own eyes."

DISCOVERY OF THE SOLAR SPOTS.—The solar spots seem to have been noticed as early as 807 A.D., although the telescope was not invented until 1610, and Galileo is considered to have discovered them in the following year.*

NUMBER AND LOCATION.—Sometimes, but rarely, the sun's disk is clear. During a period of ten years, observations were made on 1982 days, on 372 of which there were no spots seen. As many as two hundred spots have been noticed at one time. They are mostly found in two belts, one on each side of the equator, within not less than 10° nor more than 30° of latitude. They seem to herd together,—the length of the straggling group being generally parallel to the equator.

SIZE OF THE SPOTS.—It is not uncommon to find a spot with a surface larger than that of the earth. Schröter measured one more than 29,000 miles in diameter. Sir J. W. Herschel calculated that one which he saw was 50,000 miles in diameter. In 1843, one was seen which was 75,000 miles across, and was visible to the naked eye for an entire week.† On the day of the eclipse in 1858, a spot over 108,000 miles broad was distinctly seen, and attracted general attention in this country. In 1839, Captain Davis saw one which he computed was 180,000 miles long, and had an area of twenty-four billion square miles.

If these are deep openings in the luminous atmos-

---

* We read in the log-book of the good ship Richard of Arundell, on a voyage, in 1590, to the coast of Guinea, that " on the 7, at the going downe of the sunne, we saw a great black spot in the sunne ; and the 8 day, both at rising and setting, we saw the like, —which spot to me seeming was about the bignesse of a shilling, being in 5 degrees of altitude, and still there came a great billow out of the souther board."

† 1″ on the sun's surface = 450.3 miles. This spot was 2′47″ across (Schwabe).

phere of the sun, what an abyss must that be at "the bottom of which our earth could lie like a boulder in the crater of a volcano!"

SPOTS CONSIST OF DISTINCT PARTS.—From the accompanying representation, it will be seen that the spots generally consist of one or more dark portions called the *umbra*, and around that a grayish portion styled the *penumbra* (*pene*, almost, and *umbra*, black). Sometimes, however, umbræ appear without a penumbra, and vice

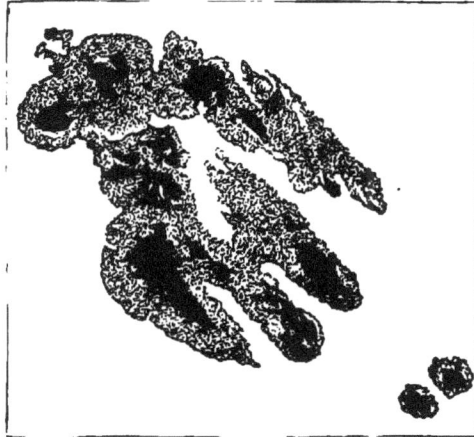

Fig. 10.

Sun-Spots.

versa. The umbra itself has generally a dense black center, called the *nucleus*. Besides this, the umbra is sometimes divided by luminous bridges.

SPOTS ARE IN MOTION.—The spots change from day to day; but all have a common movement. About fourteen days are required for a spot to pass across the disk of the sun from the eastern side, or *limb*, to the western; in fourteen days, it reappears, changed in form perhaps, but generally recognizable.

SPOTS APPARENTLY CHANGE THEIR SPEED AND FORM AS THEY PASS ACROSS THE DISK.—A spot is seen on the eastern limb; day by day it progresses, with a gradually-increasing rapidity, until it reaches the center ; it then slowly loses its rapidity, and finally disappears on the western limb. The diagram illustrates the apparent change which takes place in the form. Suppose at first the spot is of an oval shape ; as it approaches the center it apparently widens and becomes circular. Having passed that point, it becomes more and more oval until it disappears.

Fig. 11.

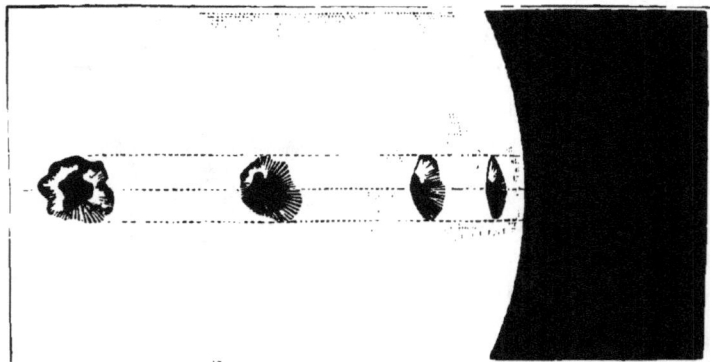

Change in Spots as they Cross the Disk.

THIS CHANGE IN THE SPOTS PROVES THE SUN'S ROTATION ON ITS AXIS.—These changes can be accounted for only on the supposition that the sun rotates on its axis : indeed, they are the precise effects which the laws of perspective demand in that case. About twenty-seven days elapse from the appearance of a spot on the eastern limb before it is seen a second time. During this period the earth has gone forward in its orbit, so that the location of the observer is changed ; allowing for this, the sun's time of rotation at the equator is about twenty-five days (25 d., 8 h., 10 m. : *Langier*).

Curiously enough, the equatorial regions move more rapidly, and complete a rotation in less time, than the rest of the sûn. While a spot near the equator performs a rotation in twenty-seven days, one situated half-way to either pole, requires nearly twenty-eight days.

SYNODIC AND SIDE-REAL REVOLUTIONS OF THE SPOTS.—We can easily under-stand why we make an allowance for the motion of the earth in its orbit. Suppose a solar spot at *a*, on a line pass-ing from the center of the earth to the center of the sun. For the spot to pass around the sun and come into that same

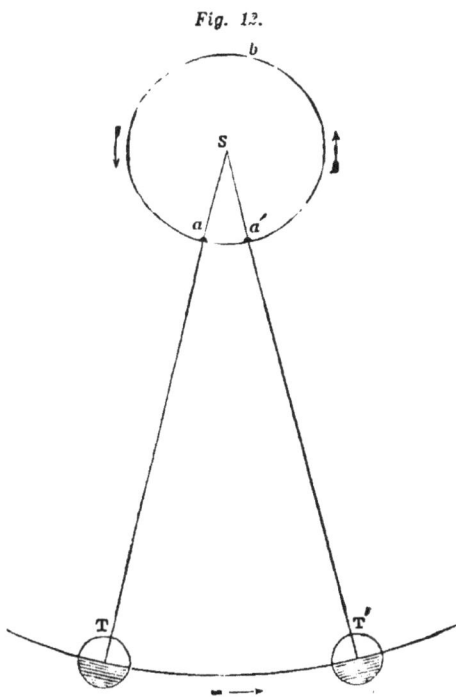

Fig. 12.

Synodic and Sidereal Revolutions.

position again, requires about twenty-seven days. But, during this time, the earth has passed on from T to T'. The spot has not only traveled around to *a* again, but also beyond that to *a'*, or the distance from *a* to *a'* more than an entire revolution. To do this, requires about two days. A revolution from *a* around to *a'* is called a *synodic*, and one from *a* around to *a* again is called a *sidereal*, revolution.

SPOTS DO NOT ALWAYS MOVE IN STRAIGHT LINES.—
Sometimes their path curves toward the north, and

Fig. 13.

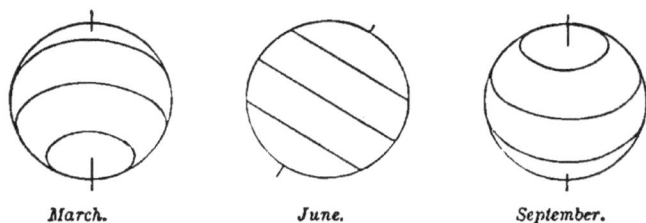

March.   June.   September.

sometimes toward the south, as in the figure. This
can be explained only on the supposition that the
sun's axis is inclined to the ecliptic (7° 15').

SPOTS HAVE A MOTION OF THEIR OWN.—Besides the
motion already named as assigned to the sun's rotation, nearly every spot seems to have an individual
motion. Some spots circle about in small elliptical
paths, often quite regularly for weeks and even
months. Immense cyclones occasionally pass over
the surface with fearful rapidity, producing rotation
and sudden changes in the spots. At other times,
however, the spots seem "to set sail and move
across the disk of the sun like gondolas over a silver
sea."

SPOTS CHANGE THEIR REAL FORM.—Spots break out
and then disappear under the eye of the astronomer.
Wollaston saw one that seemed to be shattered like
a fragment of ice when it is thrown on a frozen
surface, breaking into pieces, and sliding off in
every direction. Sometimes one divides itself into
several nuclei, while again several nuclei combine

into a single nucleus. Occasionally a spot will remain for six or eight rotations, while often it will last scarcely half an hour. Sir W. Herschel relates

Fig. 14.

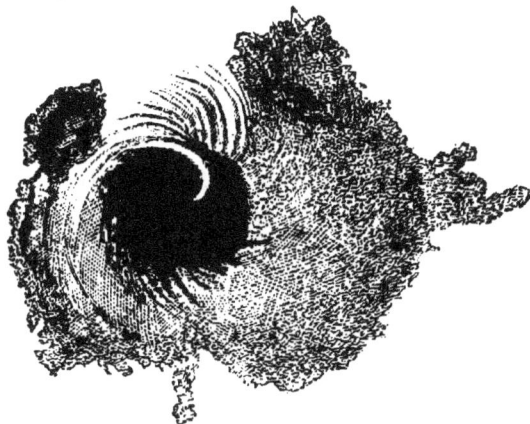

Solar Cyclone, May 5th, 1857. (Secchi.)

that, when examining a spot through his telescope, he turned away for a moment, and on looking back it was gone.

APPEARANCE OF THE SPOTS IS PERIODICAL.*—It is a remarkable fact that the number of spots increases and diminishes through a regular interval of about 11.11 years. These periodic variations are closely connected with similar variations in the aurora and magnetic earth-currents which interfere with the telegraph.

ARE THE SPOTS INFLUENCED BY THE PLANETS ?—

* The regular increase and diminution in the number of the spots was discovered by Schwabe of Prussia, who watched the sun so carefully that it is said "for thirty years the sun never appeared above the horizon without being confronted by his imperturbable telescope."

Many astronomers of high standing believe that the solar spots are especially sensitive to the approach of Mercury and Venus, on account of their nearness, and of Jupiter, because of its size ; that the area of the spots exposed to view from the earth is uniformly greatest when any two of the larger planets come into line with the sun ; and that when both Venus and Jupiter are on the side of the sun opposite to us, the spots are much larger than when Venus alone is in that position. Most authorities, however, doubt the accuracy of these observations, and deny this planetary influence altogether.

SPOTS DO NOT INFLUENCE FRUITFULNESS OF THE SEASON.—Herschel first advanced the idea that years of abundant spots would be years also of plentiful harvest. This is not now generally received. What two years could be more dissimilar than 1859 and 1860 ? Both abounded in solar spots, yet, in Europe, one was a fruitful year and the other one of almost famine. Whether the spots influence the weather is still a mooted question.

SPOTS ARE COOLER THAN THE SURROUNDING SURFACE.—It seems that the breaking out of a spot sensibly diminishes the temperature of that portion of the sun's disk. The faculæ, on the other hand, do not increase the temperature (*Secchi*).

SPOTS ARE DEPRESSIONS. — Careful observations show that, in general, the "floor," so to speak, of the umbra is sunk from two to six thousand miles below the level of the luminous surface (*Young*).

COMPARATIVE BRIGHTNESS OF SPOTS AND SUN.— If we represent the ordinary brightness of the

Fig. 15.

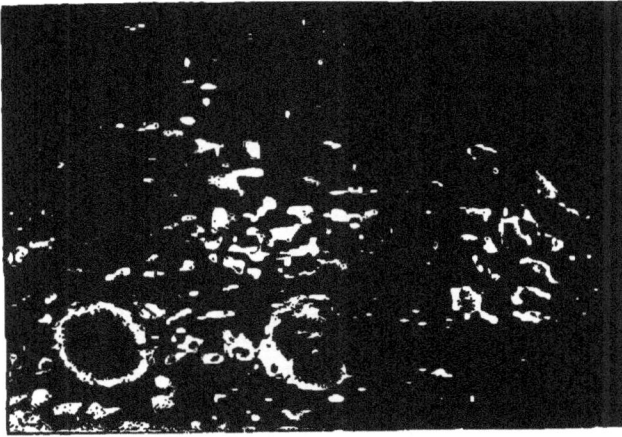

Photographic View of Spots and Faculæ.

sun by 1,000, then that of the penumbra would be about 800, and that of the umbra, 540 (*Langley*). There may be much light and heat radiated by a spot, which seems black as compared with the sun ; for we remember that even a calcium light, held between our eyes and the sun, appears as a black spot on the disk of that luminary.

Fig. 16.

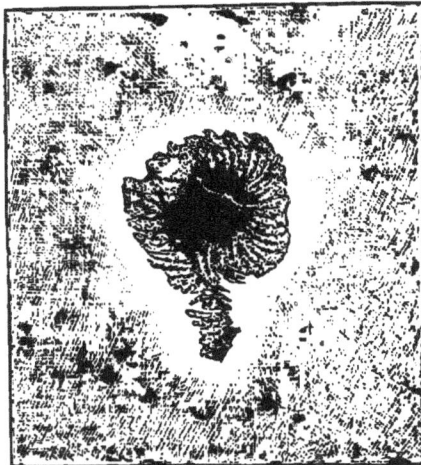

Faculæ.

APPEARANCE OF THE SUN'S SURFACE.—Even a telescope of moderate power will show the surface of the sun to have a peculiar *mottled* appear-

3

ance not unlike that of an orange skin. But, under favorable circumstances and with a telescope of high power, the solar disk is found to be covered with small, intensely bright bodies irregularly distributed.

Fig. 17.

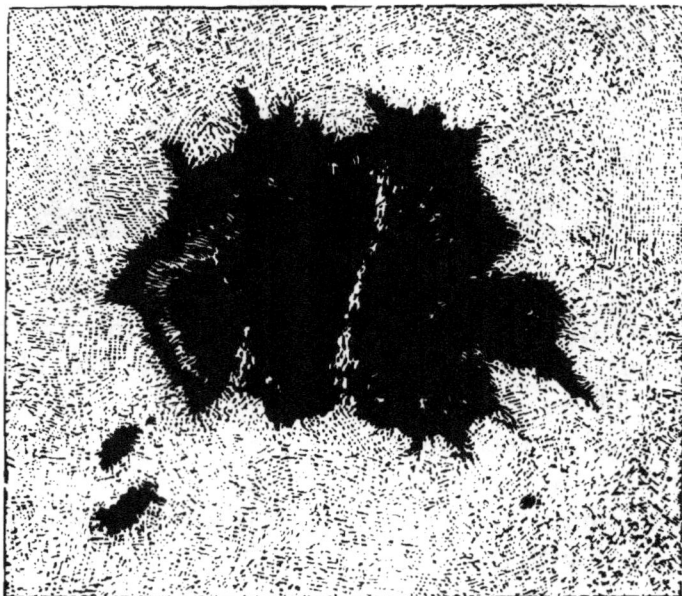

Willow-Leaf.

These are now known as *rice-grains.** They are often apparently crowded together in luminous ridges, or streaks, termed *faculæ* (*facula*, a torch) ; while the rice-grains themselves, according to Prof. Langley, are composed of *granules.* Minute as a

* Various observers describe the solar surface differently. A peculiar, elongated, leaf-shaped appearance of the rice-grains, called the willow-leaf structure, is shown in Fig. 17, as seen by Nasmyth. Newcomb compares the sun's appearance to that of a plate of rice-soup. Young says it frequently resembles bits of straw lying parallel to one another—the " thatched-straw formation."

Typical Sun-spot, of Dec. 1873, showing the filaments pointing to the center.

granule seems, probably the smallest has a diameter of, at least, 100 miles.

Physical Constitution of the Sun.*—Of the constitution of the sun, and the cause of the solar spots, very little is definitely known.

WILSON'S THEORY supposed that the sun is composed of a solid, dark globe, surrounded by three atmospheres. The first, nearest the black body of the sun, is a dense, cloudy covering, possessing high reflecting power. The second is called the *photosphere*. It consists of an incandescent gas, and is the seat of the light and heat of the sun, being the sun that we see. The third, or outer one, is transparent—very like our atmosphere.

According to this theory, the spots are to be explained in the following manner. They are simply openings in these atmospheres made by powerful upward currents. At the bottom of these chasms, we see the dark sun as a *nucleus* at the center, and around this the cloudy atmosphere—the *penumbra*. This explains a black spot with its penumbra. Sometimes the opening in the photosphere may be smaller than that in the inner or cloudy atmosphere ; in that case there will be a black spot without a penumbra.

It will be natural to suppose that when the heated gas of the photosphere, or second atmosphere, is violently rent asunder by an eruption or current from below, luminous ridges will be formed by the heaped-up gas on every side of the opening. This would account for the *faculæ* surrounding the sun-

* For the views of various authorities on the constitution of the sun, solar spots, etc., see Newcomb's Astronomy, third edition, p. 271.

spots. It will be natural, also, to suppose that sometimes the cloudy atmosphere below will close up first
over the dark surface of the sun, leaving only an
opening through the photosphere, disclosing at the
bottom a grayish surface of *penumbra*. We can

Fig. 19.

Wilson's Theory.

readily see, also, how, as the sun revolving on its
axis brings a spot nearer and nearer to the center,
thus giving us a more direct view of the opening, we
can see more and more of the dark body. Then as
it passes by the center the nucleus will disappear,

until finally we can see only the side of the fissure, the penumbra, which, in its turn, will vanish.

THE PRESENT THEORY* is deduced from the results of Spectrum Analysis, of which we shall hereafter speak. It is constantly being modified by new discoveries. But we may, in general, believe the sun to be a vast, fiery body, surrounded by an atmosphere of substances volatilized by the intense heat. Among these, we recognize familiar elements, as iron, copper, &c.

The different portions of the sun are thought to be arranged thus : (1). The *nucleus,* probably gaseous ; † (2). The *photosphere,* an envelope several thousand miles thick, which constitutes the visible part of the sun ; (3). The *chromosphere,* composed of luminous gas, mostly hydrogen, and the seat of enormous *protuberances,* tongues of fire, which dart forth, sometimes at the rate of 150 miles per second, and to a distance of over 100,000 miles ; (4). The *corona,*‡ an outer appendage of faint, pearly light, consisting of streamers reaching out often several hundred thousand miles. Of these solar constituents, the eye and the telescope ordinarily reveal only the photosphere ; the rest are seen during a total eclipse or by means of the spectroscope.

The outer portion of the sun radiates its heat and

---

* As Kirchhoff, by his discoveries in Spectrum Analysis, laid the foundation of this theory, it is often called after him.

† The interior of the sun, if gaseous, must be powerfully condensed, because of the tremendous pressure of the atmosphere. The high temperature, however, prevents the gas from liquefying. The rain-storms on the sun, if such ever occur, consist of drops of molten iron, copper, zinc, &c., vaporized by the enormous heat ; and often a tempest would drive before it this white-hot, metallic blast, with a speed of 100 miles per second.

‡ This is so called because, during a total eclipse, it forms around the moon a corona, or glory, that is the most wonderful feature of this rare event. (See p. 141.)

light, and, becoming cooler, sinks ; the hotter matter in the interior then risęs to take its place, and thus convection currents are established (Physics, p. 193). The cooler, descending currents are darker, and the hotter, ascending ones are lighter ; this gives rise to the mottled look of the sun. At times, this occurs on a grand scale, and the heated, up-rushing masses form the faculæ, and the cooler, down-rushing ones produce the solar spots.

The Heat of the Sun is generally considered to be produced by condensation, whereby the size of the sun is constantly decreasing, and its potential energy thus converted into kinetic. The dynamic theory accounts for the heat and the solar spots by assuming that there are vast numbers of meteors revolving around the sun, and that these constantly rain down upon the surface of that luminary.* Their motion, thus stopped, is changed to heat, and feeds this great central fire. Were Mercury to strike the sun in this way, it would generate sufficient heat to compensate the loss by radiation for seven years.

Doubtless, the solar heat is gradually diminishing, and will ultimately be exhausted. In time, the sun will cease to shine, as the earth did long since. Newcomb says that in 5,000,000 years, at the present rate, the sun will have shrunk to half its present size, and that it cannot sustain life on the earth more than 10,000,000 years longer. Of this we may be assured, there is enough to support life on our globe for millions of years yet to come.

* The heat of the sun could be maintained by an annual contraction of 220 feet in its diameter, a decrease so insignificant as to be imperceptible with the best instruments ; or by the annual impact of meteors equal in amount to ⅟ the mass of Mercury.

II.—THE PLANETS.

INTRODUCTION.

The Planets will be described in regular order, passing outward from the sun. In this journey, we shall examine each planet in turn, noticing its distance, size, length of year, duration of day and night, temperature, climate, number of moons, and other interesting facts, showing how much we can know of its world-life in spite of its wonderful distance. We shall encounter the earth in our imaginary wanderings through space, and shall explain many celestial phenomena already partially familiar to us.

In all these worlds, we shall find traces of the same Divine hand, molding and directing in conformity to one universal plan. We shall discover that the laws of light and heat are invariable, and that the force of gravity, which causes a stone to fall to the ground, acts similarly upon the most distant planet. Even the elements of which the planets are composed will be familiar to us, so that a book of natural science published here might, in its general features, answer for use in a school on Mars or Jupiter.

Common Characteristics (*Hind*). — 1. The planets move in the same direction around the sun ; their course, as viewed from the north side of the ecliptic, being contrary to the motion of the hands of a watch.

2. They describe elliptical paths around the sun,— not differing much from circles.

3. Their orbits are more or less inclined to the ecliptic, and intersect it in two points—the nodes,— one-half of the orbit lying north, and the other south of the earth's path.

4. They are opaque bodies, and shine by reflecting the light they receive from the sun.

5. They rotate upon their axes in the same way as the earth. This we know by telescopic observation to be the case with many planets, and by analogy the rule may be extended to all. Hence, they have the alternation of day and night.

6. Agreeably to the principles of gravitation, their velocity is greatest at that part of their orbit nearest the sun. and least at that part most distant from it : in other words, they move quickest in *perihelion*, and slowest in *aphelion*.

Comparison of the two Groups of the Major Planets. (*Chambers.*)—Separating the major planets into two groups, if we take Mercury, Venus, the Earth, and Mars as belonging to the interior, and Jupiter, Saturn, Uranus, and Neptune to the exterior group, we shall find that they differ in the following respects :

1. The interior planets. with the exception of the Earth and Mars, are not attended by any satellite, while all the exterior planets have satellites.

2. The average density of the first group consider-

ably exceeds that of the second, the approximate ratio being 5 : 1.

3. The mean duration of the axial rotations, or the mean length of the day of the interior planets, is much longer than that of the exterior ; the average in the former case being about twenty-four hours, but in the latter only about ten hours.

Properties of the Ellipse.—In Fig. 20, S and S' are the *foci* of the ellipse ; A C is the *major axis ;* B D, the *minor* or *conjugate axis ;* O, the *center :* or, astronomically, O A is the *semi-axis-major* or mean

Fig. 20.

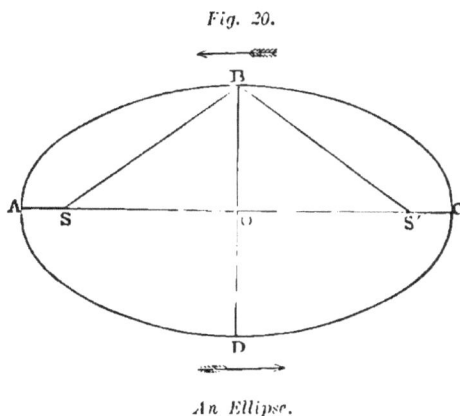

An Ellipse.

distance, O B the *semi-axis-minor :* the ratio of O S to O A is the *eccentricity ;* the least distance, S A, is the *perihelion distance ;* the greatest distance, S C, the *aphelion distance.*

Characteristics of a Planetary Orbit.—It will not be difficult to follow in the mind the additional characteristics of a planet's orbit. Take two hoops, and bind them into an oval shape. Incline one slightly to the other, as shown in Fig. 21.   Let the horizontal hoop represent the ecliptic.   Imagine a planet following the inclined hoop, or ellipse ; at a certain point it rises above the level of the ecliptic :* this point is called the *ascending node*, and the op-

Fig. 21.

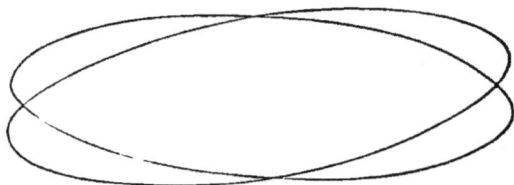

Planetary Orbits.

posite point of intersection is termed the *descending node*.   A line connecting the two nodes is the *line of the nodes*.   The *longitude of the node* is its distance from the first point of Aries, measured on the ecliptic, eastward.

Comparative Size of Planets *(Chambers).*—The following scheme will assist in obtaining some notion of the magnitude of the planetary system.   Choose a level field or common ; on it place a globe two feet in diameter for the Sun : Vulcan will then be represented by a small pin's head, at a distance of about twenty-seven feet from the center of the ideal sun ; Mercury by a mustard-seed, at a distance of eighty-two feet ; Venus by a pea, at a distance of 142 feet ; the Earth, also, by a pea, at a distance of 215 feet ; Mars by a small pepper-corn, at a distance of 327 feet ; the minor planets by grains of sand, at distances varying from 500 to 600 feet.   If space will permit, we may place a moderate-sized

---

* Lockyer beautifully says : "We may imagine the earth floating around the sun on a boundless ocean, both sun and earth being half immersed in it.   This level, this plane, the *plane of the ecliptic* (because all eclipses occur in it), is used by astronomers as we use the sea-level.   We say a mountain is so far above the level of the sea.   The astronomer says the star is so high above the level of the ecliptic.

orange nearly one-quarter of a mile distant from the starting point to represent Jupiter ; a small orange two-fifths of a mile for Saturn ; a full-sized cherry three-quarters of a mile distant for Uranus ; and lastly, a plum 1¼ miles off for Neptune, the most distant planet yet known.   Extending

Fig. 22.

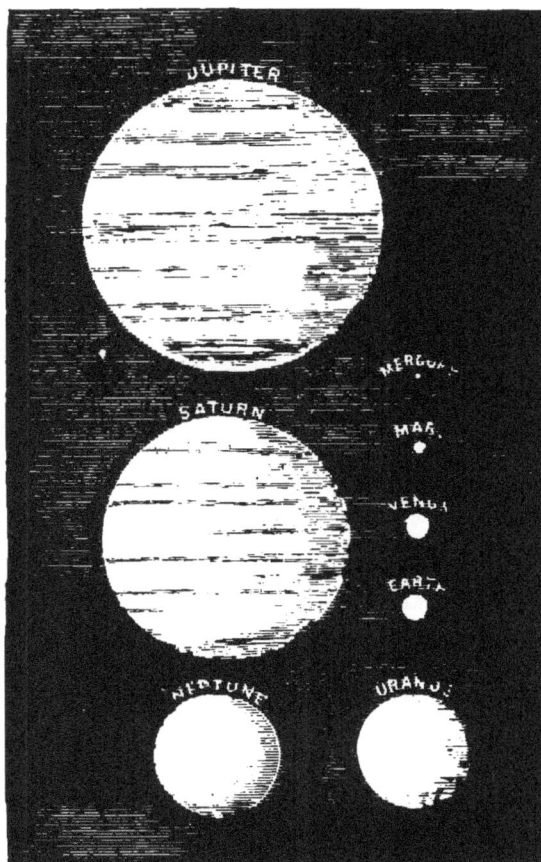

Comparative Size of the Planets.

this scheme, we should find that the aphelion distance of Encke's comet would be at 880 feet ; the aphelion distance of Donati's comet of 1858 at six miles ; and the nearest fixed star at 7,500 miles.

According to this scale, the daily motion of Vulcan in its orbit would be 4⅔ feet; of Mercury, 3 feet; of Venus, 2 feet; of the Earth, 1⅝ feet; of Mars, 1½ feet; of Jupiter, 10½ inches; of Saturn, 7½ inches; of Uranus, 5 inches; and of Neptune, 4 inches. This illustrates the fact that the orbital velocity of a planet decreases as its distance from the sun increases.[*]

Conjunction of Planets.—The grouping together of two or more planets within a limited area of the heavens is a rare event. The earliest record we have is the one of Chinese origin (p. 6), stating that a conjunction of Mars, Jupiter, Saturn, and Mercury

Fig. 23.

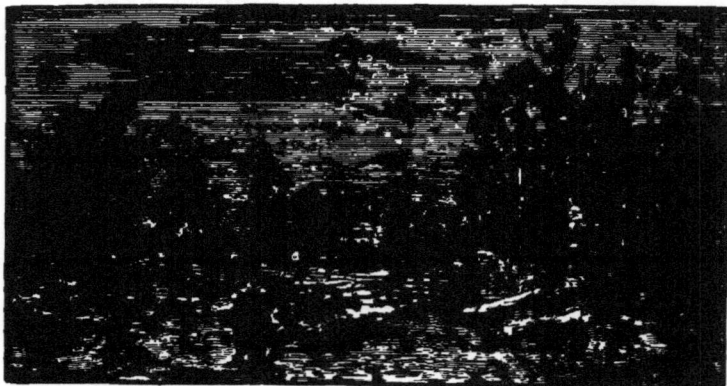

Venus and Jupiter in Conjunction, January 30, 1868.

occurred in the reign of the Emperor Chuenhio. Astronomers tell us that this took place Feb. 28, 2446 B. C., between 10° and 18° of Pisces. There is a very general impression, however, that this conjunction was afterward calculated and chronicled in their records. In 1725, Venus, Mercury, Jupiter, and

[*] If we accept the Nebular Hypothesis (p. 255), we can easily understand the reason of this; the exterior planets, being made earlier, had the motion of the nebula during its earlier stage. The rotation-velocity of the nebula kept increasing, and so, of course, each planet possessed a higher rate of orbital speed than the preceding one.

Mars appeared in the same field of the telescope. In 1859, Venus and Jupiter came so near each other that they appeared to the naked eye as one object.

Are the Planets Inhabited?—This question is one which very naturally arises, when we think of the planets as worlds in so many respects similar to our own. We can give no satisfactory answer. Many think that the only object God can have in making a world is to form an abode for man. Our own earth was evidently fitted up, although perhaps not created, for this express purpose. Everywhere about us we find proofs of special forethought and adaptation. Coal and oil in the earth for fuel and light, forests for timber, metals in the mountains for machinery, rivers for navigation. and level plains for corn. The human body, the air, light, and heat are all fitted to one another with exquisite nicety.

When we turn to the planets, we do not know but God has other races of intelligent beings who inhabit them, or even entirely different ends to attain. Of this, however, we are assured, that, if inhabited, the conditions on which life is supported vary much from those familiar to us. When we come to speak of the different planets, we shall see (1) how they differ in light and heat, from seven times our usual temperature to less than $\frac{1}{1000}$; (2) in the intensity of the force of gravity, from $2\frac{1}{2}$ times that of the earth to less than $\frac{1}{2}$; (3) in the constitution of the planet itself, from a density $\frac{1}{2}$ heavier than that of the earth to one nearly that of cork.

The temperature may often sweep downward

through a scale of 2,000° in passing from Mercury
to Neptune. No human being could reside on the
former, while we cannot conceive of any polar inhab-

Fig. 24.

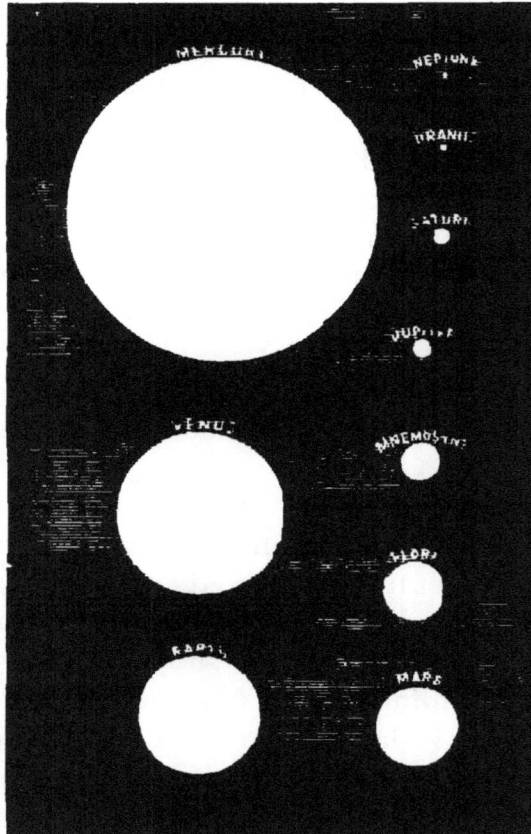

Size of the Sun as seen from the Planets.

itant who could endure the intense cold of the latter.
At the sun, one of our pounds would weigh over 27
pounds; on our moon, the pound weight would be-

come only about two ounces; while on Vesta, one of the planetoids, a man could easily spring sixty feet in the air and sustain no shock in falling. Yet, while we speak of these peculiarities, we do not know what modification of the atmosphere or physical features may exist on Mercury to temper the heat, or on Neptune to reduce the cold.

With all these diversities, we must, however, admit the power of an all-wise Creator to form beings adapted to the life and the land, however different from our own. The Power that prepared a world for us, could as easily and perfectly prepare one for other races. May it not be that the same love of diversity, that will not make two leaves after the same pattern nor two pebbles of the same size, delights in worlds peopled by races as diverse? *

While, then, we cannot affirm that the planets are inhabited, analogy would lead us to think that they are, and that the most distant star that shines in the arch of heaven may give light and heat to living beings under the care and government of Him who enlivens the densest forest with the hum of insects, and populates even a drop of water with its teeming millions of animalcules.

Divisions of the Planets.—The planets are divided into two classes: (1). *Inferior*, or those whose orbits are within that of the earth—viz., Mercury, Venus; (2). *Superior*, or those whose orbits are beyond that

---

* Astronomers conceive the universe to contain worlds in every possible stage of development, from the primary, gaseous nebula, to a worn-out, dead globe, like the moon. At a certain period in its existence, each world may be fitted to support life. Millions may now be in that condition; others may be approaching, while others have passed it.

of the earth—viz., Mars, Jupiter, Saturn, Uranus, Neptune.

Motions of a Planet as seen from the Sun.—Could we stand at the sun and watch the movements of the planets, they would all be seen revolving with different velocities in the order of the zodiacal signs. But to us, standing on one of the planets, itself in motion, the effect is changed. To an observer at the sun all the motions would be real, while to us many are only apparent. The position of a planet, as seen from the center of the sun, is called its *heliocentric place ;* as seen from the center of the earth, its *geocentric place.* When Venus is at inferior conjunction, an observer at the sun would see it in the opposite part of the heavens from that in which it would appear to him if viewed from the earth.

Motions of an Inferior Planet.—An inferior planet is never seen by us in any part of the sky opposite to the sun at the time of observation. It cannot recede from him as much as 90°, or ¼ the circumference, since it moves in an orbit entirely enclosed by the orbit of the earth. Twice in every revolution it is in conjunction (☌) with the sun,—an *inferior conjunction* (A) when it comes between the earth and the sun, and a *superior conjunction* (B) when the sun lies between it and the earth.

When the planet attains its greatest distance east or west (as we see it) from the sun, it is said to be at its *greatest elongation.*

When passing from B to A it is east of the sun, and from A to B it is west of the sun. When east of the sun, it sets later than the sun, and hence is

evening star : when west of the sun, it rises earlier
than the sun, and hence is *morning star.* An inferior
˙planet is never visible when in *superior* conjunction,
as its light is then lost in the greater brilliancy of
the sun.   When in *inferior* conjunction, it some-

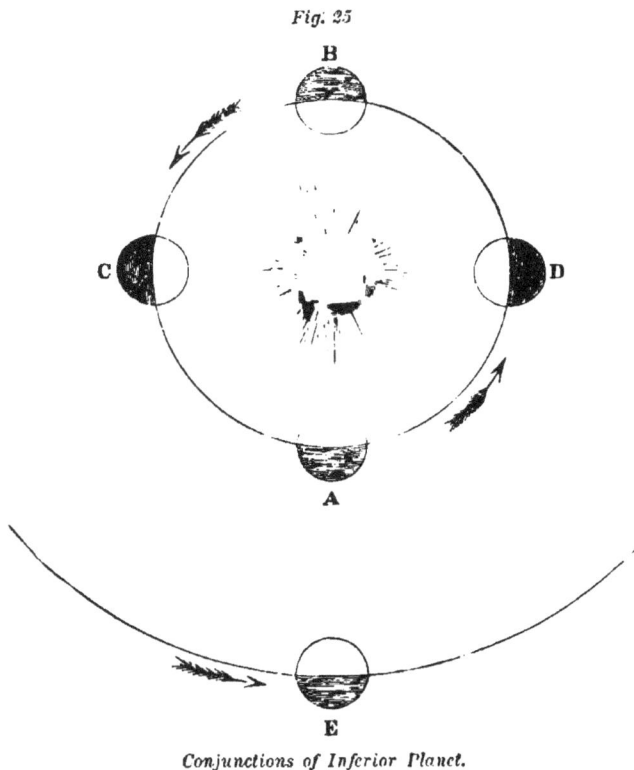

Fig. 25

Conjunctions of Inferior Planet.

times passes in front of the sun, and appears to us as
a round, black spot swiftly moving across his disk.
This is called a *transit.*

RETROGRADE MOTION OF AN INFERIOR PLANET.—
Suppose the earth at A (Fig. 26), and the planet at B.

Now, while the earth is passing to F, the planet will pass to D,—the arc AF being shorter than BD, because the nearer a planet is to the sun the greater its velocity. While the planet is at B, we locate it at C on the ecliptic, in Gemini; but at D, it appears to us to be at G, in Taurus. So that the planet has

Fig. 26.

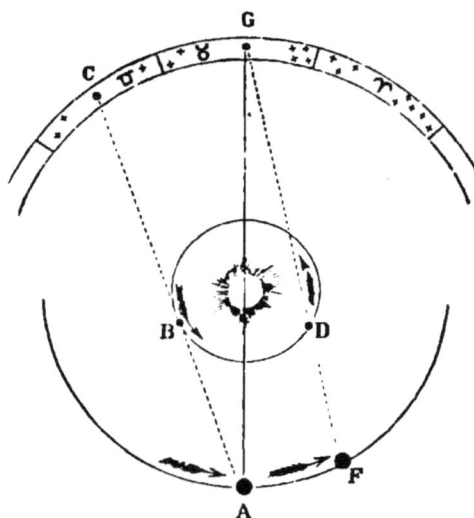

Retrograde Motion.

retrograded through an entire sign on the ecliptic, while its course all the while has been directly forward in the order of the signs; and to an observer at the sun, such would have been its motion.

PHASES OF AN INFERIOR PLANET. — An inferior planet presents all the phases of the moon. At superior conjunction, the whole illumined disk is turned toward us; but the planet is lost in the sun's rays:

therefore neither Mercury nor Venus ever presents a complete circular appearance, like the full moon. A little before or after superior conjunction, an inferior planet may be seen with a telescope ; but the whole of the light side is not turned toward us, and so the planet appears *gibbous,* like the moon between the first quarter and full. At its greatest elongation, the planet shows us only one-half its illumined disk ; this decreases, becoming more and more crescent toward inferior conjunction, at which time the un-illumined side is toward us.

Fig. 27.

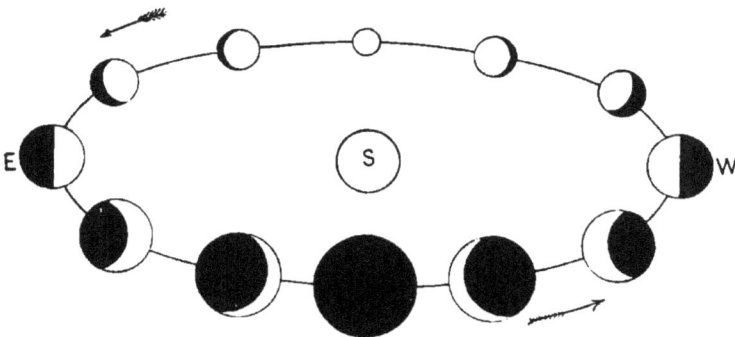

Phases of an Inferior Planet.

Motions of a Superior Planet.—The superior planet moves in an orbit which entirely surrounds that of the earth. When the earth is at E (Fig. 28), the planet at L is said to be in *opposition* to the sun (☍). It is then at its greatest distance from him—180°. The planet is on the meridian at midnight, while the sun is on the corresponding meridian on the opposite side of the earth ; or the planet may be rising, when the sun is just setting. When the planet is at N, it is in *conjunction*, and being lost in the sun's rays is invisible to us.   When 90° east or west of the sun, the planet is said to be in *quadrature* (□).

RETROGRADE MOTION OF A SUPERIOR PLANET.—Suppose the earth to be at E and the planet at L, and that we move on to G while the planet passes on to

Fig. 23.

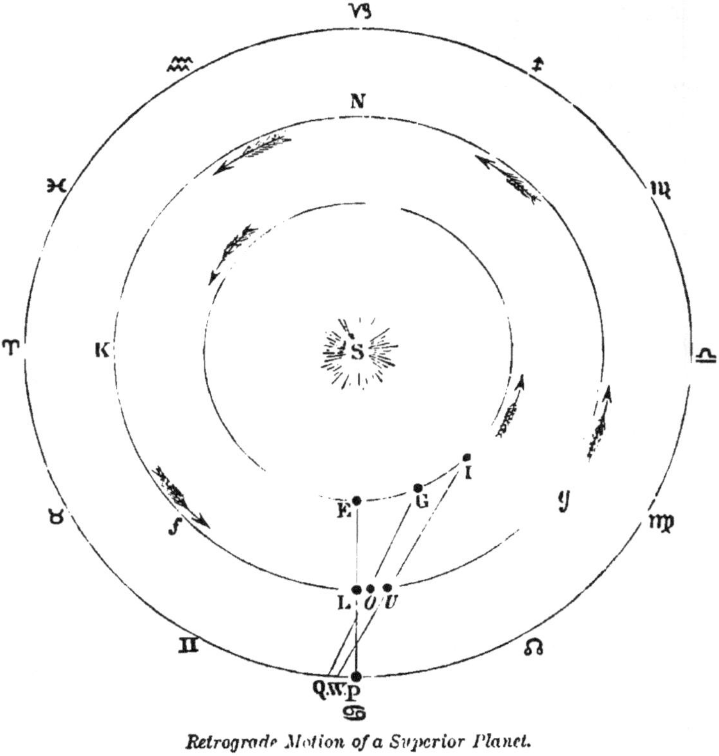

Retrograde Motion of a Superior Planet.

O—the distance EG being longer than LO, the reverse of what takes place in the movements of the inferior planets ; at E, we should locate the planet at P on the ecliptic, in the sign Cancer; but at G, it would appear to us at Q, in the sign Gemini, having

apparently retrograded on the ecliptic the distance
PQ, while it was all the time moving on in the
direct order of the signs. Now, suppose the earth
passes on to I and the planet to U, we should then
see it at the point W, further on in the ecliptic than
Q, which indicates direct motion again, and at some
point near Q the planet must have appeared without
motion.

After this, it will continue direct until the earth has
completed a large portion of her orbit, as we can
easily see by imagining various positions of the earth
and planet, and then drawing lines as we have just
done, noticing whether they indicate direct or retrograde motion. The greater the distance of a planet
the less it will retrograde, as we can perceive by
drawing another orbit outside the one represented in
the cut, and making the same suppositions concerning it as those we have already explained.

Sidereal and Synodic Revolution.—The interval
of time required by a planet to perform a revolution
from one fixed star back to it again, is termed a
sidereal revolution (*sidus*, a star).

1. The interval of time between two similar conjunctions of an inferior planet with the earth and
the sun is termed a *synodic revolution*. Were the
earth at rest, there would be no difference between a
sidereal and a synodic revolution, and the planet
would come into conjunction twice in each revolution. Since, however, the earth is in motion, it follows that, after the planet has completed its sidereal
revolution, it must overtake the earth before they
can both come again into the same position with

regard to the sun. The faster a planet moves, **the** sooner it can do this. Mercury, traveling at a greater speed and on an inner orbit, accomplishes it much more quickly than Venus. The synodic period always exceeds the sidereal.

2. The interval between two successive conjunctions or oppositions of a superior planet is also termed a *synodic revolution.* Since the earth moves so much faster than any superior planet. it follows that, after it has completed a sidereal revolution, it must overtake the planet before they can again come into the same position with regard to the sun. The slower the planet, the sooner this can be done. Uranus, making a sidereal revolution in eighty-four years, can be overtaken more quickly than Mars, which makes one in less than two years. It consequently requires over a second revolution for the earth to catch up with Mars, only $\frac{1}{11}$ of a second one to overtake Jupiter, and but little over $\frac{1}{100}$ of a second one to come up with Uranus.

Planets as Evening and Morning Stars.—The inferior planets are evening stars from superior to inferior conjunction ; and the superior planets, from opposition to conjunction. During the other part of their revolutions, they are morning stars.

Mercury is evening star.,			about 2 months.
Venus	"	"	"	$9\frac{1}{2}$	"
Mars	"	"	"	13	"
Jupiter	"	"	"	$6\frac{1}{2}$	"
Saturn	"	"	"	6	"
Uranus	"	"	"	6	"

I. VULCAN (hypothetical).

Supposed Discovery.—Le Verrier, having detected an error in the assumed motion of Mercury, suggested, in the autumn of 1859, that there might be an interior planet, which was the cause of this disturbance. On this being made public, M. Lescarbault, a French physician and an amateur astronomer, stated that on March 26 of that year he had seen a dark body pass across the sun's disk, which might have been the unknown planet. Le Verrier visited him, and found his instruments rough and home-made, but singularly accurate. His clock was a simple pendulum, consisting of an ivory ball hanging from a nail by a silk thread. His observations were on prescription paper, covered with grease and laudanum. His calculations were chalked on a board, which he planed off to make room for fresh ones. Le Verrier became satisfied that a new planet had been discovered by this enthusiastic observer, and congratulated him upon his deserved success.

On March 20, 1862, Mr. Lummis, of Manchester, England, noticed a rapidly-moving, dark spot, apparently the transit of an inner planet. During the total eclipse of July 29, 1878, Professor Watson, of Ann Arbor Observatory, and Dr. Lewis Swift, of Rochester, claimed to have seen two Intra-Mercurial planets. As yet, however, the existence of the planet is not generally conceded. The name Vulcan and the sign of a hammer have been given to it. Its distance from the sun has been estimated at 13,000,000 miles, and its periodic time (its year) at twenty days.

II. MERCURY.

The fleetest of the gods. Sign, ☿, his wand.

Description.—Mercury is nearest to the sun of any of the definitely-known planets. When the sky is very clear, we may sometimes see it, just after sunset, as a bright, sparkling star, near the western horizon. Its elevation increases evening by evening, but never exceeds 28°.* If we watch it closely, we shall find that the planet again approaches the sun and becomes lost in his rays. Some days afterward, just before sunrise, we can see the same planet in the east, rising higher each morning, until its greatest elevation equals that which it before attained in the west. Thus the planet appears slowly but steadily to oscillate like a pendulum, to and fro, from one side to the other of the sun. The ancients, deceived by this puzzling movement, failed to discover the identity of the two stars, and called the morning star Apollo, the god of day, and the evening star Mercury, the god of thieves, who walk to and fro in the night-time seeking plunder.†

On account of the nearness of Mercury to the sun, it is difficult to be detected.‡ It is said that Copernicus, an old man of seventy, lamented in his last moments that, much as he had tried, he had never been able to see it. In our latitude and climate, we can generally easily find it if we watch for it at the time of its greatest elongation, as commonly given in the almanac.

Motion in Space.—Mercury revolves around the sun at a mean distance of about 36,000,000 miles. Its

---

* This distance varies much, owing to the eccentricity of Mercury's orbit.

† The Greeks gave to Mercury the additional name of "The Sparkling One." The astrologists looked upon it as the malignant planet. The chemists, because of its extreme swiftness, applied the name to quicksilver. The most ancient account that we have of this planet is given by Ptolemy, in his Almagest; he states its location on the 15th of November, 265 B. C. The Chinese also state that on June 9, 118 A. D., it was near the Beehive, a cluster of stars in Cancer. Astronomers tell us that, according to the best calculations, it was at that date within less than 1° of that group.

‡ An old English writer by the name of Goad, in 1686, humorously termed this planet, "A squinting lacquey of the sun, who seldom shows his head in these parts, as if he were in debt."

orbit is the most eccentric (flattened) of any among
the eight principal planets, so that, although when in
perihelion it approaches to within about 28,000,000
miles, in aphelion it speeds away 15,000,000 miles
further, or to the distance of over 43,000,000 miles.
Being so near the sun, its motion in its orbit is correspondingly rapid,—viz., thirty miles per second.*

The Mercurial year comprises only about eighty-
eight days, or nearly three of our months. Mercury
is thought to rotate upon its axis in about the same
time as the earth, so that the length of the Mercurial
day is nearly the same as that of the terrestrial
one.

Though Mercury thus completes a sidereal revolution around the sun in eighty-eight days, yet to pass
from one inferior or superior conjunction to the next
(a synodic revolution) requires 116 days. The reason
of this is, that when Mercury comes around again
to the point of its last conjunction, the earth has
gone forward, and it requires twenty-eight days for
the planet to overtake us.

The Distance from the Earth varies still more
than the distance from the sun. At inferior conjunction, Mercury is between the earth and the sun,
and its distance from us is the *difference* between
the distance of the earth and of the planet from the
sun : at superior conjunction, it is the *sum* of these
distances. Its apparent diameter in these different
positions varies in the same proportion as the distance, or nearly three to one. The greatest and least

---

* At this rate of speed, we could cross the Atlantic Ocean in two minutes.

distances vary as either planet happens to be in aphelion or perihelion.*

Dimensions.—Mercury is about 3,000 miles in *diameter*. Its *volume* is about $\frac{1}{20}$ that of the earth— *i. e.*, it would require twenty globes as large as Mercury to make one the size of the earth, or 25,000,000 to equal the sun. It is $\frac{1}{3}$ denser than the earth, its *mass* is nearly $\frac{1}{16}$ that of the earth, and a stone let drop upon its surface would fall $7\frac{1}{2}$ feet the first second. Its *specific gravity* is not far from that of tin. A pound weight removed to Mercury would weigh only about seven ounces.

Seasons.—As Mercury's axis is much inclined from a perpendicular (perhaps 70°), its seasons are peculiar. There are no distinct frigid zones; but large regions near the poles have six weeks of continuous day and torrid heat, alternating with a night of equal length and arctic cold. The sun shines perpendicularly upon the torrid zone only at the equinoxes, while he sinks far toward the southern horizon at one solstice, and as far toward the northern horizon at the other.† The equatorial regions, therefore, during each revolution, are modified in their temperature from torrid to temperate, and the tropical heat is experienced alternately toward the north and the south of what we call the temperate zones.

There is no marked distinction of zones as with us, but each zone changes its character twice during the

---

* If at inferior conjunction Mercury is in aphelion and the earth in perihelion, its distance from us is only 91,500,000 − 43,000,000 = 48,500,000 miles. If at superior conjunction Mercury is in aphelion and the earth in aphelion also, its distance from us is 94,500,000 + 43,000,000 = 137,500,000 miles.

† Read a chapter entitled "The Fiery World," in Proctor's Poetry of Astronomy.

Mercurial year, or eight times during the terrestrial one. An inhabitant of Mercury must be accustomed to sudden and violent vicissitudes of temperature. At one time, the sun not only thus pours down its vertical rays, and in a few weeks after sinks far toward the horizon, but, on account of Mercury's

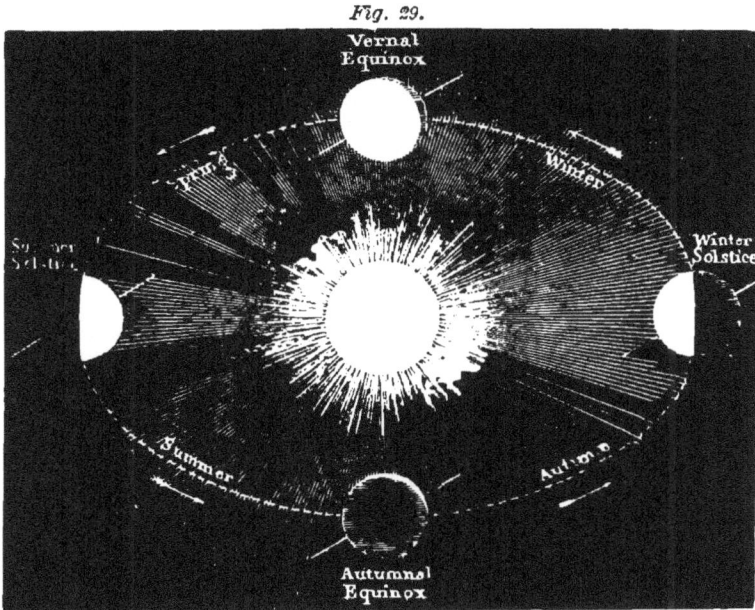

Orbit and Seasons of Mercury.

elliptical orbit, when in perihelion the planet approaches so near the sun that the heat and light are ten times as great as ours, while in aphelion it recedes so as to reduce the amount to four and a half times. The average heat is about seven times that of the earth,—a temperature sufficient to turn water into steam, and even to melt zinc.

The relative length of the days and nights is much more variable than with us. The sun, apparently seven times as large as it seems to us, must be a magnificent spectacle, and illumine every object with insufferable brilliancy. The evening sky is, however, lighted by no moon.

Telescopic Features.—Through the telescope, Mercury presents all the phases of the moon, from a slender crescent to gibbous, after which its light is lost in that of the sun. These phases prove that Mercury is spherical, and shines by the light reflected from the sun. Being an inferior planet, we never see it when full, and hence the brightest, nor when nearest the earth, as then its dark side is turned toward us.

Owing to the dazzling light, and the vapors almost always hanging around our horizon, this planet has not of late received much attention; the data here given are mainly based upon the observations of the older astronomers, and are, therefore, not universally accepted. Mercury is thought by some to have a dense, cloudy atmosphere, that materially diminishes the intensity of its heat and, perhaps, makes it habitable, though others assert that the atmosphere is too insignificant to be detected. Some dark bands about the planet's equator indicate, perhaps, an equatorial zone. There are, also, lofty heights which intercept the light of the sun, and deep valleys plunged in shade. One mountain is claimed to be over eleven miles high, or about $\frac{1}{250}$ the diameter of the planet.*

---

* The height of the loftiest peak of the Himalayas is only 29,000 feet, or about $\frac{1}{1430}$ part of the earth's diameter.

III. VENUS.

The Queen of Beauty. Sign ♀, a looking-glass.

Description.—Venus, the next in order to Mercury, is the most brilliant of the planets.* She presents the same appearances as Mercury. Owing, however, to the larger size of her orbit, her greatest apparent oscillations are nearly 48° east and west of the sun,† or about 20° more than those of Mercury. She is therefore seen much earlier in the morning and much later at night. She is morning star from inferior to superior conjunction, and evening star from superior to inferior conjunction.

Venus is the most brilliant about five weeks before and after inferior conjunction, at which time the planet is bright enough to cast a shadow at night. If, in addition, at this time of greatest brilliancy, Venus is at or near her highest north latitude, she may be seen with the naked eye in full daylight.‡ This occurs once in eight years—the interval required for the earth and planet to return to the same situation in their orbits ; eight complete revolutions of the

* When visible before sunrise, she was called by the ancients Phosphorus, Lucifer, or the Morning Star, and when she shone in the evening after sunset, Hesperus, Vesper, or the Evening Star.

† This distance varies only about 3°, owing to the slight eccentricity of Venus's orbit.

‡ Arago relates that Buonaparte, upon repairing to the Luxembourg, when the Directory was about to give him a *fête*, was much surprised at seeing the multitude paying more attention to the heavens above the palace than to him or his brilliant staff. Upon inquiry, he learned that these curious persons were observing with astonishment a star which they supposed to be that of the Conqueror of Italy. The emperor himself was not indifferent when his piercing eye caught the clear lustre of Venus smiling upon him at midday.

earth about the sun occupying nearly the same time as thirteen of Venus.

Motion in Space.—Venus has an orbit the most nearly circular of any of the principal planets. Her mean distance from the sun is about 67,000,000 miles, which varies at aphelion and perihelion 1,000,000 miles,—a contrast to Mercury, which varies 15,000,000 miles.

Venus makes a complete revolution around the sun in about 225 days, at the mean rate of twenty-two miles per second; hence her year is equal to about seven and one-half of our months. This is a *sidereal* revolution, as it would appear to an observer at the sun; a *synodic* revolution requires 584 days.

Mercury, we remember, catches up with the earth in twenty-eight days after it reaches the point where it left the earth at the last inferior conjunction. But it takes Venus nearly two and a half revolutions to overtake the earth and come into the same conjunction again. This grows out of the fact that she has a longer orbit than Mercury, and moves only about one-sixth faster than the earth, while Mercury travels nearly twice as fast as our planet. Venus rotates upon her axis in about twenty-four hours: so the length of her day does not differ essentially from ours.

Distance from the Earth.—Like that of Mercury, the distance of Venus from the earth, when in inferior conjunction, is the difference between the distances of the two planets from the sun; when in superior conjunction, the sum of these distances.

When nearest to us, Venus is only about 25,000,000 miles away.

Figure 30 represents her apparent dimensions at the extreme, mean, and least distances from us. The variation is nearly as the numbers 10, 18, and 65. It would be natural to think that the planet is the brightest when the nearest, and thus the largest, but

Fig. 30.

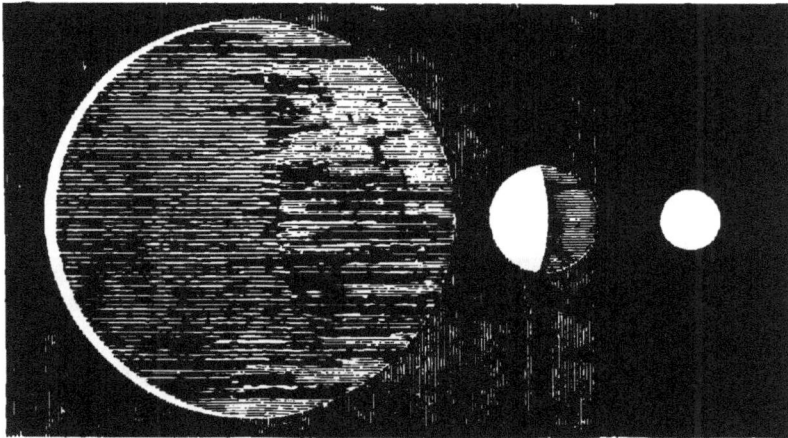

Extreme, Mean, and Least Apparent Size of Venus; and her Phases.

we should remember that then the bright side is toward the sun, and the unillumined side toward us. Indeed, at the period of greatest brilliancy, of which we have spoken, only about *one-fourth* of her light is visible. At this time, however, observers have noticed the entire contour of the planet to be of a dull gray hue, as seen in the cut.

Dimensions.—Venus is about 7,600 miles in *diameter*. The *volume* and *density* of the planet are each about nine-tenths that of the earth. A stone let fall upon her surface would fall fourteen feet in the first second : a pound weight removed to her equator would weigh about fourteen ounces. From this we see that the force of gravity does not decrease exactly in proportion to the size of the planet, any more than it increases with the size of the sun. The reason is, that the body is brought nearer the mass of the small planet, and so feels its attraction more fully than when far out upon the circumference of a large body,—the attraction increasing as the square of the distance from the particles decreases.

Seasons.—Since the axis of Venus is very much inclined from a perpendicular, her seasons are similar

Fig. 31.

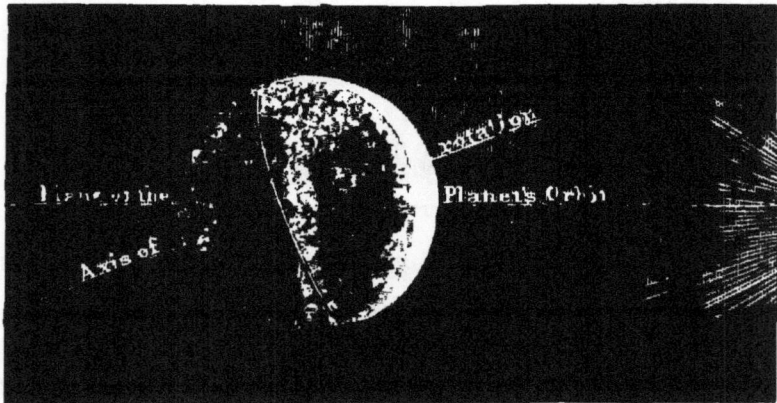

Venus at her Solstice.

to those of Mercury. The torrid and temperate zones overlap each other, and the polar regions have, alternately, at one solstice a torrid temperature, and at the other a prolonged arctic cold. The inequality of the nights is very marked. The heat and light are double that of the earth, while the circular form of her orbit gives nearly an equal length to her four seasons.

If the inclination of her axis is 75°, as some astronomers hold, her tropics must be 75° from the equator, and her polar circles 75° from the poles. The torrid zone is, therefore, 150° in width. The torrid and frigid zones interlap through a space of 60 , midway between the equator and the poles.

Telescopic Features.—Venus, being an interior planet, presents, like Mercury, all the phases of the moon.*

She is thought to have a dense, cloudy atmosphere. This was suggested by the fact that at the transit of

Fig. 32.

Crescent and Spots of Venus.

Venus over the sun in 1761, 1769, and 1882, a faint ring of light surrounded the black disk of the planet.

---

* This was discovered by Galileo, and was among the first achievements of his telescopic observations. It had been argued against the Copernican system that, if true, Venus should wax and wane like the moon. Indeed, Copernicus himself boldly declared that, if means of seeing the planets more distinctly were ever invented, Venus would be found to present such phases. Galileo, with his telescope, proved this fact, and thus vindicated the Copernican theory.

The evidence of an atmosphere, as well as of mountains, however, rests upon the peculiar appearance attending her crescent shape.

1. The luminous part does not end abruptly; on the contrary, the light diminishes gradually. This diminution can be explained by a twilight caused by an atmosphere which diffuses the rays of light into regions of the planet where the sun is already set. Thus, on Venus, as on the earth, the evenings are lighted by twilight, and the mornings by dawn.

2. The edge of the enlightened portion of the planet is uneven and irregular. This appearance is doubtless the effect of shadows cast by mountains.

Spots have been noticed on her disk which are considered to be traceable to clouds. Herschel thinks that we never see the body of the planet, but only her atmosphere loaded with vapors, which may mitigate the glare of the intense sunshine.

Satellites.—Venus is not known to have any moon.

IV. THE EARTH.

Sign, ⊕, a circle with Equator and Meridian.

THE EARTH is the next planet we meet in passing outward from the sun. To the beginner, it seems strange enough to class our world among the heavenly bodies. *They* are brilliant, while *it* is dark and opaque; they appear light and airy, while it is solid and firm; we see in it no motion, while they are constantly changing their position ; they seem mere points in the sky, while it is vast and extended.

Yet, at the very beginning, we are to consider the

Fig. 33.

The Earth in Space.

earth as a planet shining brightly in the heavens, and appearing to other worlds as a planet does to us.

We are to learn that it is in motion, flying through its orbit with inconceivable velocity ; that it is not fixed, but hangs in space, held by an invisible power of gravitation which it cannot evade ;* that it is small and insignificant beside the mighty globes that so gently shine upon us in the far-off sky ; that, in fact, it is only one atom in a universe of worlds, all firm and solid, and all, perhaps, equally fitted to be the abode of life.

Dimensions.—The earth is not "round like a ball," but flattened at the poles. Its form is that of an oblate spheroid. Its polar diameter is about 7,899 miles, and its equatorial about 7,925¼. The compression is, therefore, 26¼ miles. (See table in Appendix.) If we represent the earth by a globe one yard in diameter, the polar diameter would be one-tenth of an inch too long. The circumference of the earth is nearly 25,000 miles. Its density is about 5½ times that of water. Its weight is 6,069,000,000,000,000,000,000 tons.

The inequalities of the earth's surface, arising from valleys, mountains, etc., have been likened to the roughness on the rind of an orange. On a globe sixteen inches in diameter, the land, to be in proportion, should be represented by the thinnest writing paper, the hills by very fine grains of sand, and elevated ranges by thick drawing-paper. To represent the deepest wells or mines, a scratch should be made that would be invisible except with a glass.

---

* Were the sun's attractive force upon the earth replaced by the largest steel telegraph wire, it would require nine wires for each square inch of the sunward side of our globe, to hold the earth in her orbit.

The Rotundity of the Earth is proved in various ways: (1) By the fact that vessels have sailed around the earth;* (2) when a ship is coming into port, we see the masts first; (3) the shadow of the earth on the moon is circular; (4) the polar star seems higher in the heavens as we pass north; and (5) the horizon expands as we ascend an eminence.† If we climb to the top of a hill, we can see further than when on the plain at its foot. Our eyesight is not improved; it is only because ordinarily the curvature of the earth shuts off the view of distant objects, but when we ascend to a higher point, we can see further over the side of the earth. The curvature is eight inches per mile, $2^2 \times 8^{in} = 32$ inches for two miles, $3^2 \times 8^{in}$ for three miles, etc. An object of these respective heights would be just hidden at these distances.

Apparent and Real Motion.—In endeavoring to understand the various appearances of the heavenly bodies, it is well to remember how in daily life we

* It is curious, in connection with this well-known fact, to recall the arguments urged by the Spanish philosophers against the reasoning of Columbus, when he assured them that he could arrive at Asia just as certainly by sailing west as east. "How," they asked, "can the earth be round? If it were, then on the opposite side the rain would fall upward, trees would grow with their branches down, and everything would be topsy-turvy. Every object on its surface would certainly fall off, and if a ship by sailing west should get around there, it would never be able to climb up the side of the earth and get back again. How can a ship sail up hill?"

† "The history of aëronautic adventure affords a curious illustration of this same principle. The late Mr. Sadler, the celebrated aëronaut, ascended on one occasion in a balloon from Dublin, and was wafted across the Irish Channel, when, on his approach to the Welsh coast, the balloon descended nearly to the surface of the sea. By this time the sun was set, and the shades of evening began to close in. He threw out nearly all his ballast, and suddenly sprang upward to a great height, and by so doing brought his horizon to *dip* below the sun, producing the whole phenomenon of a western sunrise. Subsequently descending in Wales, he, of course, witnessed a second sunset on the same evening."

transfer motion. On the cars, when in rapid movement, the fences and the trees seem to glide by us,
while we sit still and see them pass. On a bridge,
when we are at rest, we watch the undulations of
the waves, until at last we come to think that they
are stationary and we are sweeping up the stream.

"In the cabin of a large vessel going smoothly before the wind on still
water, or drawn along a canal, not the smallest indication acquaints us
with the 'way it is making.' We read, sit, walk, as if we were on land.
If we throw a ball into the air, it falls back into our hand ; if we drop it,
it alights at our feet. Insects buzz around us as in the free air, and smoke
ascends in the same manner as it would do in an apartment on shore. If,
indeed, we come on deck, the case is in some respects different ; the air,
not being carried along with us, drifts away smoke and other light bodies,
such as feathers cast upon it, apparently in the opposite direction to that
of the ship's progress ; but in reality they remain at rest, and we leave
them behind in the air. And what is the earth itself but the good ship we
are sailing in through the universe, bound round the sun ; and as we sit
here in one of the 'berths,' we are unconscious of there being any 'way'
at all upon the vessel. On deck, too, out in the open air, it's all the same
so long as we keep our eyes on the ship ; but immediately we look over
the sides—and the horizon is but the 'gunwale' of our vessel—we see the
blue tide of the great ocean around us go drifting by the ship, and sparkling with its million stars as the waters of the sea itself sparkle at night
between the tropics."

Diurnal Rotation of the Earth around its Axis.
—The earth, in constantly turning from west to east,
elevates our horizon above the stars on the west, and
depresses it below the stars on the east. As the
horizon appears to us to be stationary, we assign the
motion to the stars, thinking those on the west,
which it passes over and hides, to have sunk below it,
or *set ;* and imagining those on the east, below which

it has dropped, to have moved above it, or *risen.* So, also, the horizon is depressed below the sun, and we call it *sunrise ;* it is elevated above the sun, and we call it *sunset.*

We thus see that the diurnal movement of the sun by day and the stars by night is an optical illusion, —that here as elsewhere we simply transfer motion. This seems easy enough for us to understand ; but it was the "stone of stumbling" to ancient astronomers for thousands of years. Copernicus himself, it is said, first thought of the true solution while riding on a vessel and noticing how he insensibly transferred the movement of the ship to the objects on the shore. How much grander the beautiful simplicity of this system than the cumbersome complexity of the old Ptolemaic belief !

DIURNAL MOTION OF THE SUN.—The explanation just given illustrates the apparent motion of the sun, and the cause of day and night. Suppose S to be the sun. The earth, E, turning upon its axis EF from

Fig. 34.

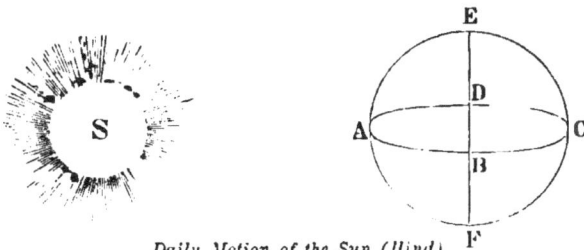

Daily Motion of the Sun (Hind).

west to east, has only half its surface illuminated at one time by the sun. To a person at D, the sun is in the horizon and day commences, that luminary appearing to rise higher and higher, with a westerly motion, as the observer is carried forward easterly by the earth's diurnal rotation to A, where he has the sun in his meridian, and it is consequently noon. The sun then begins to decline in the sky until the spectator arrives at B, where it sets, or is again in the horizon on the west side, and night begins. He moves on to C, which marks his position at midnight, the sun being then on the meridian of places on the opposite part of the earth, and he is brought round again to D, the point of sunrise, when another day commences.

UNEQUAL RATE OF DIURNAL MOTION.—Different points upon the surface of the earth revolve with different velocities. At the poles the speed of rotation is nothing, while at the equator it is greatest, or over 1,000 miles per hour. At Quito, the circle of latitude is much longer than the one at the mouth of the St. Lawrence, and the velocities vary in the same proportion. The former place moves at the rate of about 1,038 miles per hour; the latter, 682 miles. In our latitude (41°) the speed is about 780 miles per hour. We do not perceive this wonderful velocity with which we are flying through the ether, because the atmosphere moves with us.*

Were the earth suddenly to stop its rotation, the terrible shock would, without doubt, destroy the

---

* "An ingenious inventor once suggested that we should utilize the earth's rotation, as the most simple and economical, as well as rapid mode of locomotion that could be conceived. This was to be accomplished by rising in a balloon to a height inaccessible to aerial currents. The balloon, remaining immovable in this calm region, would simply await the moment when the earth, rotating underneath, should present the place of destination to the eyes of travelers who would then descend. A well-regulated watch and an exact knowledge of longitude would thus render traveling possible from

entire race of man; while we, with houses, trees, rocks, and even the oceans, would be hurled, in one confused mass, headlong into space. On the other hand, were the rate of rotation to increase, the length of the day would be proportionately shortened, and the weight of all bodies decreased by the centrifugal force thus produced. If the rotary movement should become swift enough to reduce the day to eighty-four minutes, the force of gravity would be overcome, and, at the equator, all bodies would be without weight; if the speed were still further increased, loose bodies would fly off from the earth like water from a swiftly-turned grindstone, while we should be compelled constantly to "hold on" to avoid sharing the same fate.* But against such a catastrophe we are assured by the immutability of God's laws. "He is the same yesterday, to-day, and forever."

UNEQUAL DIURNAL ORBITS OF THE STARS.—In figure 35, let O represent our position on the earth's surface; E Z B, our meridian; E I B K, our horizon : P and P', the north and south poles; Z, the zenith;

---

east to west, all voyages north or south being interdicted. This suggestion has only one fault; it supposes that the atmospheric strata do not revolve with the earth. Upon that hypothesis, since we rotate (at London) with the velocity of 333 yards in a second, there would result a wind in the contrary direction ten times more violent than the most terrible hurricane. Is not the absence of such a state of things a convincing proof of the participation of the atmospheric envelope in the general movement?"— GUILLEMIN.

* Laplace concluded in 1799 that the inequalities of the earth's rotation were too insignificant for measurement. But, more recently, Delaunay has shown from the moon's acceleration that a minute change, caused by the friction of the sea and atmosphere upon the earth's surface, has taken place, producing a variation in the length of the day. The acceleration of the moon in its path, is, however, only seven feet per century, or less than an inch per annum, and the time of the earth's rotation has increased but $\frac{121}{10000}$ of a second in 2,400 years.—BALL.

Z'. the nadir ; and G I C K the celestial equator. Now
P B, it will be seen, is the elevation of the north pole
above the horizon, or the latitude of the place.

Fig. 35.

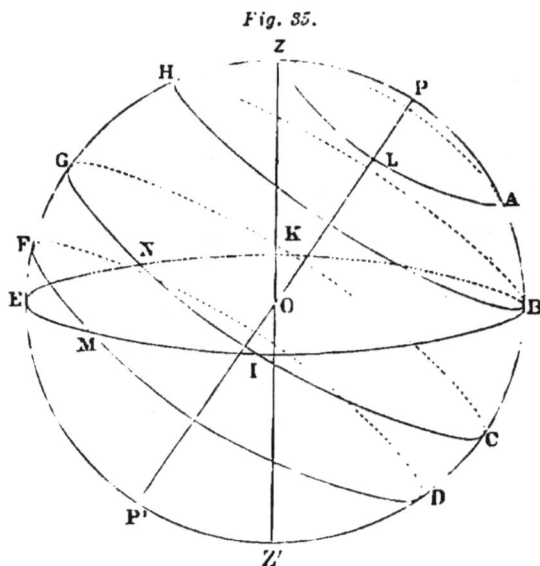

Suppose we should see a star at A. on the meridian
below the pole. The earth revolves in the direction
G I C ; the star will therefore move along A L to Z,
when it is on the meridian above the pole. It continues its course along the dotted line around to A
again, when it is on the meridian below the pole.
having made a complete circuit around the pole, but
not having descended below our horizon.

A star rising at B would just touch the horizon :
one at I would move on the celestial equator, and
would be above the horizon as long a time as it is
below,—twelve hours in each case ; a star rising at M
would come just above the horizon and set again at N.

UNEQUAL DIURNAL VELOCITIES OF THE STARS.—
The stars appear to us to be set in a concave shell
which revolves daily about the earth. As different parts of the earth *really* rotate with varying velocities, so the stars *appear* to revolve at different
rates of speed. Those near the pole, having a small
orbit, revolve very slowly, while those near the
celestial equator move at the greatest speed.

APPEARANCE OF THE STARS AT DIFFERENT PLACES
ON THE EARTH.—Were we placed at the north pole,
Polaris would be directly overhead, and the stars
would seem to pass around us in circles parallel to
the horizon, and increasing in diameter from the
upper to the lower ones. Were we placed at the
equator, the pole-star would be at the horizon, and
the stars would move in circles perpendicular to the
horizon, and decreasing in diameter, north and south
from those in the zenith, while we could see one
half of the path of each star. Were we placed in the
southern hemisphere, the circumpolar stars would
revolve about the south pole, and the others in
circles resembling those in our sky, only the points of
direction would be reversed to correspond with the
pole. Were we placed at the south pole, the appearance would be the same as at the north pole, except
that no star is there to mark the direction of the
earth's axis.

Motion of the Earth in Space about the Sun.—
The earth revolves in an elliptical path about the
sun at a mean distance of 93,000,000 miles.

The eccentricity of this path, which is greater
than that of the orbit of Venus, changes about

$\frac{4}{100.000}$ per century. The orbit would, therefore, finally become circular, were it not that, after the lapse of some thousands of years, the eccentricity will begin to increase again, and will thus vary through all time within definite, although yet undetermined limits. The circumference is nearly 600,000,000 miles, and the earth pursues this wonderful journey at the rate of over eighteen miles per second.

This revolution of the earth about the sun gives rise to various phenomena, of which we shall now proceed to speak.

1. CHANGE IN THE APPEARANCE OF THE HEAVENS IN DIFFERENT MONTHS.—In Fig. 36, suppose A B C D to be the orbit of the earth, and E F G H the sphere of the fixed stars, surrounding the sun in every direction. When our globe is at A, the stars about E are on the meridian at midnight. Being seen from the earth in the quarter opposite to the sun, they are favorably placed for observation. The stars at G, on the contrary, will be invisible, for the sun intervenes between them and the earth : they are on the meridian of the spectator about the same time as the sun, and are hidden in his rays.

In three months, the earth has passed over one-fourth of its orbit, and has arrived at B. Stars about F now appear on the meridian at midnight ; those at E, which previously occupied their places, have descended toward the west ; while those about G are just coming into sight in the east.

In three months more, the earth is situated at C,

and stars about G shine in the midnight sky, those
at F having, in their turn, vanished in the west;

Fig. 36.

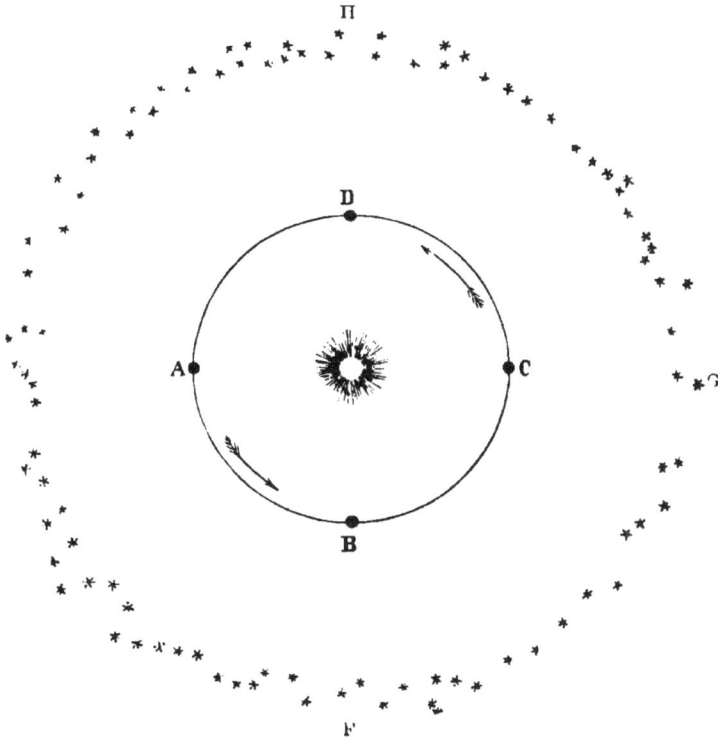

Appearance of the Heavens in Different Seasons (Hind).

stars at E are on the meridian at noon, and consequently hidden in daylight; and those about H are
just making their appearance in the east. One revolution of the earth will bring the same stars again
on the meridian at midnight.

Thus the earth's motion round the sun as a center
explains the varied aspect of the heavens in the
summer and winter skies.

2. YEARLY PATH OF THE SUN THROUGH THE HEAVENS.—We have spoken of the diurnal motion of the sun. We shall now speak of its *second* apparent motion, its yearly path among the stars,—the ecliptic.* If we look at Fig. 37, we can see how the motion of the earth in its orbit is transferred to the sun, and causes him to appear to travel in a fixed path through the heavens. When the earth is in any part of its orbit, the sun seems to us to be in the point directly opposite. For example, when the earth is in Libra (♎)†—autumnal equinox—the sun is in Aries (♈)—vernal equinox; when the sun enters the next sign, Taurus (♉), the earth has passed on to Scorpio (♏). Thus, as the earth moves through her orbit, the sun seems to pass along the opposite side of the ecliptic, making the circuit of the heavens in a year, and returning, at the end of that time, to the same place among the stars. The ecliptic crosses the celestial equator at two points, called the *equinoxes*. (See page 30).

* This yearly movement of the sun among the fixed stars is not so apparent to us as his daily motion, because his superior light blots out the stars. But if we notice a star at the western horizon just at sunset, we can tell what constellation the sun is in : then wait two or three nights, and we shall find that this star has set, and others have taken its place. Thus we can trace the sun through the year in his path among the fixed stars in the horizon.

† When we say "the earth is in Libra," we mean that a spectator placed at the sun would see the earth in that part of the heavens which is occupied by the sign Libra, while a spectator on the earth would see the sun, at the same time, in that part of the heavens which is occupied by the sign Aries. Just so, on June 21st, the earth enters Capricorn, and the sun, Cancer. It is customary, however, having reference solely to the sun's place, to locate the vernal equinox in Aries, and the autumnal equinox in Libra ; the summer solstice in Cancer, and the winter solstice in Capricorn. In figure 37, the terms "summer solstice," "autumnal equinox," etc., refer to the season upon the earth, and to the location of the sun in the ecliptic, but are not the names of those points on the earth's orbit. The zodiacal signs are inserted for convenience of illustration, to show where the earth would be located by a solar spectator ; the pupil should remember, however, that the signs belong to the ecliptic—which is the projection of the plane of the earth's orbit upon the celestial sphere, and not to the earth's path.

3. APPARENT MOVEMENT OF THE SUN, NORTH AND SOUTH.—Having now spoken of the apparent *diurnal* and *annual* motions of the sun, there yet remains a *third* motion. In summer, at midday, the sun is high in the heavens; in winter, he is low, near the southern horizon. In summer, he is a long time above the horizon; in winter, a short time. In summer, he rises and sets north of the east and west points; in winter, south of the east and west points. This subject is so intimately connected with the next, that we shall understand it best when taken in connection with that topic.

4. CHANGE OF THE SEASONS. VARIATION IN LENGTH OF DAY AND NIGHT.—By studying Fig. 37, and imagining the various positions of the earth in its orbit, let us try to understand the following points :

I. *Obliquity of the ecliptic.*—The axis of the earth is inclined $23\frac{1}{2}°$ from a perpendicular to its orbit. This angle is called the obliquity of the ecliptic.

II. *Parallelism of the axis.*—In all parts of the orbit, the axis of the earth is parallel to itself, and points almost exactly toward the North Star (p. 217).

Nature reveals to us nothing more permanent than the axis of rotation in anything that is rapidly turned. It is its rotation that keeps a boy's hoop from falling. For the same reason, a quoit retains its direction when whirled, and stays in the same plane at whatever angle it may be thrown. A man slating a roof wishes to throw a slate to the ground : he whirls it perpendicularly, and it will strike on the edge without breaking. So long as a top spins there

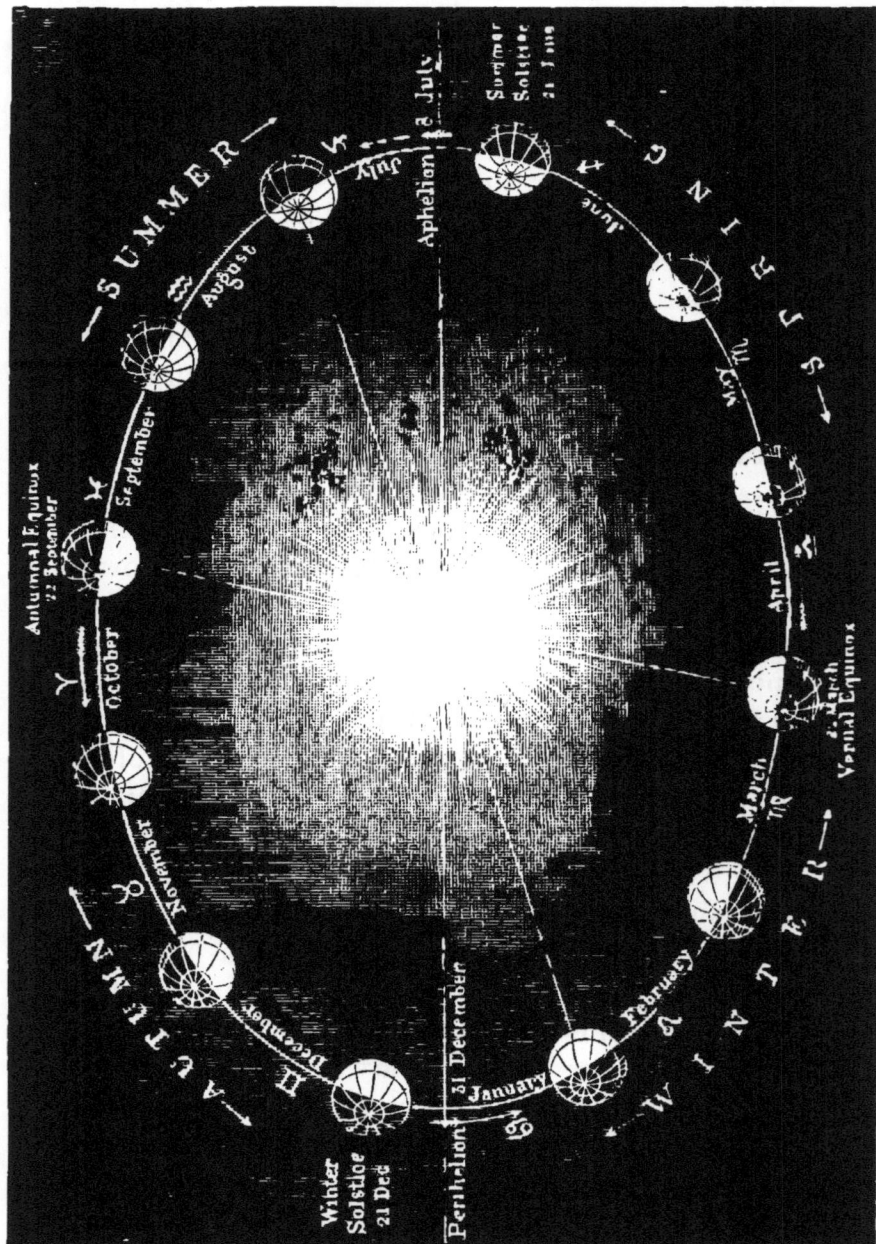

Fig. 37

The Orbit of the Earth as seen by an Observer at the Sun. (See Note, p. 94.)

is no danger of its falling, since its tendency to keep
its axis of rotation parallel is greater than the attraction of the earth. This wonderful law would lead
us to think that the axis of the earth always points
in the same direction, even if we did not know it
from direct observation.

III. *The rays of the sun strike the various portions
of the earth, when in any position, at different
angles.*—When the earth is in Libra, and also when
in Aries, the sun's rays strike vertically at the equator, and more and more obliquely in the northern
and southern hemispheres, as the distance from the
equator increases, until at the poles they strike
almost horizontally.

This variation in the direction of the rays produces a corresponding variation in the intensity of
the sun's heat and light at different places, and
accounts for the difference between the torrid and
polar regions.

IV. *As the earth changes its position the angle at
which the rays strike any portion is varied.*—Take
the earth when it is in Capricornus (♑) and the sun
in Cancer (♋). He is now overhead, $23\frac{1}{2}°$ *north* of
the equator. His rays strike less obliquely in the
northern hemisphere than when the earth was in
Libra. Let six months elapse : the earth is now in
Cancer and the sun in Capricornus ; and he is overhead, $23\frac{1}{2}°$ *south* of the equator. His rays strike less
obliquely in the southern hemisphere than before,
but in the northern hemisphere more obliquely.
These six months have changed the direction of the
sun's rays on every part of the earth's surface. This

accounts for the difference in temperature between summer and winter.*

V. *Equinoxes.*—At the equinoxes, one half of each hemisphere is illuminated : hence the name Equinox (*æquus*, equal ; and *nox*, night). At these points of the orbit, the days and nights are equal over the entire earth,† each being twelve hours in length.

VI. *Northern and southern hemispheres unequally illuminated.*—While one-half of the earth is constantly illuminated, the proportion of the northern or the southern hemisphere that is in daylight or darkness varies at all times, except at the equinoxes. When more than half of a hemisphere is in the light, its days are longer than the nights, and *vice versa.*

VII. *The seasons and the comparative length of the days and nights in the South Temperate Zone, at any time, are the reverse of those in the North Temperate Zone, except at the Equinoxes, when the days and nights are of equal length.*

VIII. *The Summer Solstice.*—At the time of the summer solstice, which occurs about the 21st of June, the sun is overhead $23\frac{1}{2}°$ north of the equator, and if his vertical rays could leave a golden line on the surface of the earth as it rotates, they would mark the Tropic of Cancer. The sun is at its furthest northern declination ; he ascends the highest he is ever seen above our horizon, and rises and sets north of the east and west points. He seems now to stand still in his northern and southern course,

---

* The long nights and short days of winter, and the short nights and long days of summer, are also important factors in producing this difference of temperature.

† Except a small space at each pole.

and hence the name *Solstice* (*sol*, the sun ; *sto*, I stand). The days in the north temperate zone are longer than the nights. It is our summer, and the 21st of June is the longest day of the year.

In the south temperate zone it is winter, and the shortest day of the year. The circle that separates day from night extends 23½° beyond the north pole, and if the sun's rays could in like manner leave a golden line on that day, they would trace on the earth the Arctic Circle. It is the noon of the long six-months polar day. The reverse is true at the Antarctic Circle, and it is there the midnight of the long six-months polar night (p. 117).

IX. *The Autumnal Equinox.*—The earth crosses the aphelion point about the 1st of July. It is then at its furthest distance from the sun, which each day rises and sets a trifle further toward the south, passing through a lower circuit in the heavens. At the time of the autumnal equinox,* the 22nd of September, he is on the equinoctial, and if his vertical rays could leave a line of golden light, they would mark on the earth the circle of the equator. It is autumn in the north temperate zone and spring in the south temperate zone. The days and nights are equal over the whole earth, the sun rising at 6 A.M. and setting at 6 P.M., exactly in the east and the west, where the equinoctial intersects the horizon.

X. *The Winter Solstice.*—The sun after passing the equinoctial—"crossing the line"—sinks lower toward the southern horizon each day. At the

* The precise time of the equinoxes and solstices varies each year, but within a small limit.

time of the winter solstice. about the 21st of December, the sun is directly overhead 23½° south of the equator, and if his vertical rays could leave a line of golden light, they would mark on the earth's surface the Tropic of Capricorn. He is at his furthest southern declination, and rises and sets south of the east and west points. It is our winter, and the 21st of December is the shortest day of the year.

In the south temperate zone it is summer, and the longest day of the year. The circle that separates day from night extends 23½° beyond the south pole, and if the sun's rays in like manner could leave a line of golden light, they would mark the Antarctic Circle. It is there the noon of the long six-months polar day. At the Arctic Circle the reverse is true ; the rays fall 23½° short of the north pole, and it is there the midnight of the long six-months polar night. Here again the sun appears to us to stand still a day or two before retracing his course, and it is therefore called the Winter Solstice.

XI. *The Vernal Equinox.*—The earth reaches its *perihelion* about the 31st of December. It is then nearest the sun, which rises and sets each day further and further north, and climbs up higher in the heavens at midday. Our days gradually increase in length, and our nights shorten in the same proportion. About the 21st of March the sun reaches the equinoctial, at the vernal equinox. He is overhead at the equator, and the days and nights are again equal. It is our spring, but in the south temperate zone it is autumn.

XII. *Yearly path finished.*—The earth moves on in

its orbit through the spring and the summer months. The sun continues his northerly course, ascending each day higher in the heavens, and his rays becoming less and less oblique. About the 21st of June, he again reaches his furthest northern declination, and is at the summer solstice.

We have thus traced the yearly path, and noticed the course of the changing seasons, with the length of the days and nights. The same series has been repeated through the ages of the past, and will be through the future till time shall be no more.

XIII. *Distance of the earth from the sun varies.*— We notice, from what we have just seen, that we are nearer the sun in winter than in summer by 3,000,000 miles. The obliqueness with which the rays strike the north temperate zone at that time prevents our receiving any special benefit from this favorable position of the earth.

XIV. *Southern summer.* — The inhabitants of the south temperate zone have their summer while the earth is in perihelion, and the sun's rays are about $\frac{1}{30}$ warmer than when in aphelion, our summer-time. This will perhaps partly account for the extreme heat of their season.[*] The southern winters, for a similar reason, are colder ; and this makes the average yearly temperature about the same as ours.

XV. *Extremes of heat and cold not at the solstices.* —We do not have our greatest heat at the time of the summer solstice, nor our greatest cold at the winter

[*] Captain Sturt, in speaking of the extreme heat of Australia, says that matches accidentally dropped on the ground were ignited. A recent official report states that, in South Australia, January, 1882, the heat, in the sun, was 180°—only 32° below the boiling-point.

solstice. After the 21st of June, the earth, already warmed by the genial spring days, continues to receive more heat from the sun by day than it radiates by night : thus its temperature still increases. On the other hand, after the 21st of December, the earth continues to become colder, because it loses more heat during the night than it receives during the day.

XVI. *Summer longer than winter.*—As the sun is not in the center of the earth's orbit, but at one of its foci, the earth, from the time of the vernal to that of the autumnal equinox, passes through more than one-half of its orbit. The summer is, therefore, longer than the winter. The difference is enhanced by the variation in the earth's velocity at aphelion and at perihelion.

XVII. *Varying velocity of earth.*—From the time of the vernal equinox until the earth passes its aphelion, the solar attraction tends to check its speed ; thence until the time of the autumnal equinox, the attraction is partly in the direction of its motion, and so increases its velocity. The same principle applies when going to and from perihelion.

XVIII. *Curious appearance of the sun at the north pole.*—"To a person standing at the north pole, the sun appears to sweep horizontally around the sky every twenty-four hours, without any perceptible variation in its distance from the horizon. It is, however, slowly rising, until, on the 21st of June, it is twenty-three degrees and twenty-eight minutes above the horizon, a little more than one-fourth of the distance to the zenith. This is the highest point it ever reaches. From this altitude, it slowly descends, its track being represented by a spiral or screw with a very fine thread ; and in the course of three months it worms its way down to the horizon, which it reaches on the 22nd of September. On this day it slowly sweeps

around the sky, with its face half hidden below the icy sea. It still continues to descend, and after it has entirely disappeared it is still so near the
horizon that it carries a bright twilight around the heavens in its daily
circuit. As the sun sinks lower and lower, this twilight grows gradually
fainter, till it fades away. December 21st, the sun is $23\frac{1}{2}°$ below the
horizon, and this is the midnight of the dark polar winter. From this
date, the sun begins to ascend, and after a time it is heralded by a faint
dawn, which circles slowly around the horizon, completing its circuit every
twenty-four hours. This dawn grows gradually brighter, and on the 22nd
of March the peaks of ice are gilded with the first level rays of the six-
months day. The bringer of this long day continues to wind his spiral
way upward, till he reaches his highest place on the 21st of June, and his
annual course is completed."

XIX. *Results, if the axis of the earth were perpendicular to the ecliptic.*—The sun would then always
appear to move through the equinoctial. He would
rise and set every day at the same points on the
horizon, and pass through the same circle in the
heavens, while the days and nights would be equal
the year round. There would be near the equator a
fierce torrid heat, while north and south the climate
would change into temperate spring, and, lastly,
into the rigors of a perpetual winter.

XX. *Results, if the equator of the earth were perpendicular to the ecliptic.*—Were this the case, to
a spectator at the equator, as the sun leaves the
vernal equinox, he would each day pass through
a smaller circle, until at the summer solstice he
would reach the north pole, when he would halt for
a time, and then slowly return in an inverse manner.

In our own latitude, the sun would make his
diurnal rotations as described, his rays shining

past the north pole further and further, until we were included in the region of perpetual day, when he would seem to wind in a spiral course up to the north pole, and then return in a descending curve to the equator.

Precession of the Equinoxes.—We have spoken of the equinoxes as if they were stationary. Over two thousand years ago, Hipparchus (see page 8) found that they are slowly falling back along the ecliptic. Modern astronomers fix the rate at about 50″ of space annually. If we mark either point in the ecliptic where the days and nights are equal over the earth—at which time the plane of the earth's equator passes exactly through the center of the sun—we shall find the sun comes back to that position the next year, 50″ (20 m. 20 s. of time) earlier. This remarkable effect is called the *Precession of the Equinoxes,* because the position of the equinoxes in any year precedes that which they occupied the year before. Since the circle of the ecliptic is divided into 360°, it follows that the time occupied by the equinoctial points in making a complete revolution at the rate of 50″.2 per year is 25,817 years.

RESULTS OF THE PRECESSION OF THE EQUINOXES.— In Fig. 37, we see that the plane of the earth's equator is inclined to that of the ecliptic. In order that the plane of the terrestrial equator should pass through the sun's center 50″ earlier, it is necessary that the plane itself should slightly change its place. The axis of the earth is always perpendicular to this plane. hence it follows that the axis is not rigorously parallel to itself. It varies in direction, so that the north pole describes a small circle in the starry vault twice 23½° in diameter.

To illustrate this, let us suppose that, after a series of years, the position of the earth's equator has changed from *e f h* to *g K l* (Fig. 38). The inclination of the axis of the earth, CP, to CQ, the pole of the ecliptic, remains unchanged ; but as it must turn

Fig. 38.

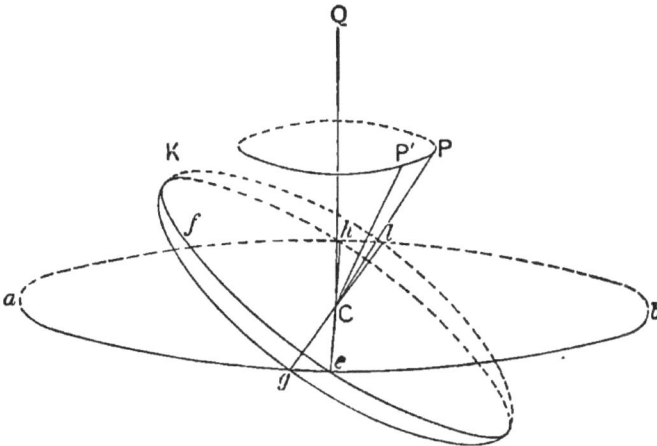

Change of Earth's Equator and Axis.*

with the equator, its position is moved from CP to CP', and the pole of the earth slowly traces the portion of a circle, PP'. The direction of this motion is the same as that of the hands of a watch, or the reverse of the revolution of the earth. The position of the north pole in the heavens is gradually but almost insensibly changing. It is now distant from the north polar star about 1½°. It will continue to approach it

* See in the Appendix a description of a simple apparatus for illustrating this subject.

until they are not more than half a degree apart.
In 12,000 years, Lyra will be our polar star : 4,500
years ago the polar star was the bright star Alpha
in the constellation Draco.   (See p. 217).

As the right ascension of the stars is reckoned
eastward from the vernal equinox along the equinoctial, the precession of the equinoxes increases the
R. A. of the stars 50″ per year. On this account, star
maps should be accompanied by the date of their
calculation, that they may be corrected to correspond with this annual variation.

The constellations of the zodiac (see p. 31) are
fixed in the heavens, while the signs are simply
abstract divisions which move with the equinox.
When named, the sun was in both the sign and
the constellation Aries, at the time of the vernal
equinox ; but since then the equinoxes have retrograded nearly a whole sign, so that now, while the
vernal equinox is in the·sign Aries, this sign corresponds to the constellation Pisces, which is therefore the first constellation in the zodiac (Fig. 86).

CAUSES OF THE PRECESSION OF THE EQUINOXES.—
Before commencing the explanation of this phenomenon, it is necessary to impress upon the mind a few
facts.   (1.) The earth is not a perfect sphere, but is
swollen at the equator.   It is like a sphere covered
with padding, increasing in thickness from the poles
to the equator ; this gives it a turnip-like shape.
(2.) The attraction of the sun is greater the nearer
a body is to it.   (3.) The attraction is not for the
earth as a mass, but for each particle separately.

In the figure, the position of the earth at the time

of the winter solstice is represented. P is the north pole ; *a b,* the plane of the ecliptic ; C, the center of the earth ; C Q, a line perpendicular to the ecliptic ; the angle Q C P, the obliquity of the ecliptic. In this position, the equatorial padding of which we have spoken—the ring of matter about the equator—is not turned exactly toward the sun, but is elevated above it. Now the attraction of the sun pulls the part D more strongly than the center ; the tendency of this

Fig. 39.

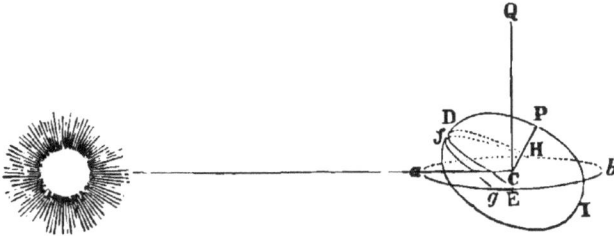

Influence of the Sun on a Mountain near the Equator.

is to bring D down to *a,* and to lift I toward *b.* The attraction for C is greater than for I, so it tends to draw C away from I, and, as at the same time D tends toward *a,* to pull I up toward *b.* The effect of this, one would think, would be to change the inclination of the axis C P toward C Q, and make it more nearly perpendicular to the ecliptic. This would be the result if the earth were not rotating upon its axis.

Let us consider the case of a mountain near the equator. This, if the sun did not act upon it, would pass through the curve H D E in the course of a semi-rotation of the earth. But, it is nearer the sun than is the center C; the attraction therefore tends to pull the mountain downward and tilt the earth over, as we have just described ; so the mountain will pass through the curve H *f g*, and, instead of crossing the ecliptic at E, will cross at *g*, a little sooner than it otherwise would. The same influence, though in a less degree, obtains on the opposite side of the earth. The mountain passes around the earth in a curve nearer to *b*, and crosses the ecliptic a little earlier.

The same reasoning will apply to each mountain and to all the protuberant mass near the equatorial regions. The final effect is slightly to turn the earth's equator so that it intersects the ecliptic sooner than it would were, it not for this attraction. At the summer solstice, the same tilting motion is produced. At the equinoxes, the plane of the earth's equator passes through the center of the sun, and therefore there is no tendency to change of position. As the axis C P must move with the equator, it slowly revolves, keeping its inclination unchanged, around C Q, the pole of the ecliptic, describing, in about 26,000 years, a small circle twice 23½° in diameter.

PRECESSION ILLUSTRATED IN THE SPINNING OF A TOP. —This motion of the earth's axis is singularly illustrated in the spinning of a top, and the more so because the forces are of an opposite character to those which act on the earth, and thus produce an opposite effect. We have seen that, if the earth had no rotation, the sun's attraction on the "padding" at the equator would bring C P nearer to C Q, but that, in consequence of this rotation, the effect really produced is that C P, the earth's axis, slowly *revolves around* C Q, the pole of the heavens, in a direction *opposite* to that of rotation.

In Fig. 40, let C P be the axis of a spinning top, and C Q the vertical line. The direct tendency of the earth's attraction is to bring C P *further from* C Q (or to make the top fall), and if the top were not spinning this would be the result; but, in consequence of the rotary motion, the inclination does not sensibly alter (until the spinning is re-

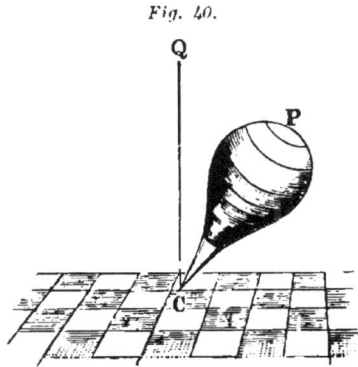

Fig. 40.

Spinning of a Top.

tarded by friction), and so C P slowly revolves around C Q in the *same direction* as that of rotation.

Nutation (*nutatio*, a nodding).—We have noticed the sun as producing precession; the moon has, however, treble his influence; for although her mass is not $\frac{1}{25,000,000}$ part that of the sun, yet she is 400 times nearer and her effect correspondingly greater.* The moon's orbit does not lie parallel to the ecliptic, but is inclined to it. Now the sun attracts the moon, and disturbs its path, as he would that of the mountain we have supposed, and the effect is the same. The intersections of the moon's orbit with the ecliptic travel backward, completing a revolution in about 18 years.

* See the Differential effect of the Sun and the Moon, under the head of the Tides.

During half of this time, the moon's orbit is inclined to the ecliptic in the same way as the earth's equator; during the other half, it is inclined in the opposite way. In the former state, the moon's attractive tendency to tilt the earth is very small, and the precession is slow; in the latter, the tendency is great, and precession goes on rapidly. The consequence of this is, that the pole of the earth is irregularly shifted, so that it travels in a slightly curved line, giving it a kind of "wabbling" or "nodding" motion, as shown —though greatly exaggerated— in Fig. 41. The obliquity of the ecliptic, which we consider $23\frac{1}{2}°$ ($23°\ 27'\ 15''$, Jan. 1, 1884. See p. 29), is the *mean* of the irregularly curved line and is represented by the dotted circle.

Fig. 41.

Path of the North Pole in the Heavens.

PERIODICAL CHANGE IN THE OBLIQUITY OF THE ECLIPTIC.—Although it is sufficiently near for all general purposes to consider the obliquity of the ecliptic invariable, yet this is not strictly the case. It is subject to a small but appreciable variation of about $46''$ per century. This is caused by a slow change of the position of the earth's orbit, due to the attraction of the planets. The effect of this movement is gradually to diminish the inclination of the earth's equator to the ecliptic (the obliquity of the ecliptic). This will continue for a time, when the angle will as gradually increase; the extreme limit of change being only $1°\ 21'$. The orbit of the earth thus vibrates backward and forward, each oscillation requiring a period of 10,000 years.

The change is so intimately blended, in its effect upon the obliquity of the ecliptic, with that caused by precession and nutation, that they are separable only in theory; in fact, they all combine to produce the waving motion we have already described. As a consequence of this variation in the obliquity of the ecliptic, the sun does not now come so far north nor decline so far south as formerly; while the position of all the terrestrial circles—the Tropics of Cancer, Capricorn, etc.—is constantly but slowly changing. As the result of this variation in the position of the orbit, some stars which were once just south of the ecliptic are now north of it, and others that were just north are now a little further north; thus the latitude of these stars is gradually changing.

CHANGE IN THE MAJOR AXIS (LINE OF APSIDES) OF THE EARTH'S ORBIT.—Besides all the changes in the position of the earth in its orbit due to precession, etc., the line connecting the aphelion and perihelion points of the orbit itself is slowly revolving. The consequence of this is a variation in the length of the seasons at different periods of time.

In the year 3958 B.C., the earth was in perihelion at the time of the autumnal equinox, so that the summer and autumn seasons were of equal length, but shorter than the winter and spring seasons, which were also equal. *

* There is much discrepancy in the views held concerning the Great Year of the astronomers, as it is often called. (See Steele's Geology, pp. 272-3, note.) The statement made in the text is that held by Lockyer, Hind, and others. The dates are those given by Chambers in his Descriptive Astronomy (3rd Edition), where the subject is fully described.

In the year 1267 A.D., the earth was in perihelion at the time of
the winter solstice, December 21, instead of January 1st, as now; the
spring quarter was therefore equal to the summer one, and the autumn
quarter to the winter one, the former being the longer.   In the year 6493
A.D., the earth will be in perihelion at the time of the vernal equinox;
summer will then be equal to autumn and winter to spring, the former
seasons being the longer.   In the year 11719 A.D., the earth will be in
perihelion at the time of the summer solstice: finally, in 16945 A.D., the
cycle will be completed, and the autumnal equinox will again coincide with
the earth's perihelion.

Permanence in the Midst of Change.—We thus
see that the ecliptic is constantly modifying its elliptical shape; that the orbit of the earth slowly oscil-
lates upward and downward; that the north pole
steadily turns its long index-finger over a dial that
marks 26,000 years; that the earth, accurately
poised in space, gently nods and bows to the attraction of sun, moon, and planets.* Thus changes are
taking place that would ultimately entirely reverse
the order of nature.   But each of these variations
has its bounds, beyond which it cannot pass.   The
promise made to man is that, "while the earth re-
maineth, seed-time and harvest, and cold and heat,
and summer and winter, and day and night shall
not cease."   The modern discoveries of astronomy
prove conclusively that the seasons are to be permanent; that the Creator, amid all these transitions,
has ordained the means of carrying out His promise
through all time.

Refraction.—The atmosphere extends above the

* These oscillations extend throughout the whole planetary system, the periods
varying from 50,000 to 2,000,000 years.   "Great clocks of eternity, which beat ages as
ours beat seconds."—*Newcomb's Astronomy, page 95.*

earth about 500 miles (Physics, p. 116). Near the surface it is dense, while in the upper regions it is

Fig. 42.

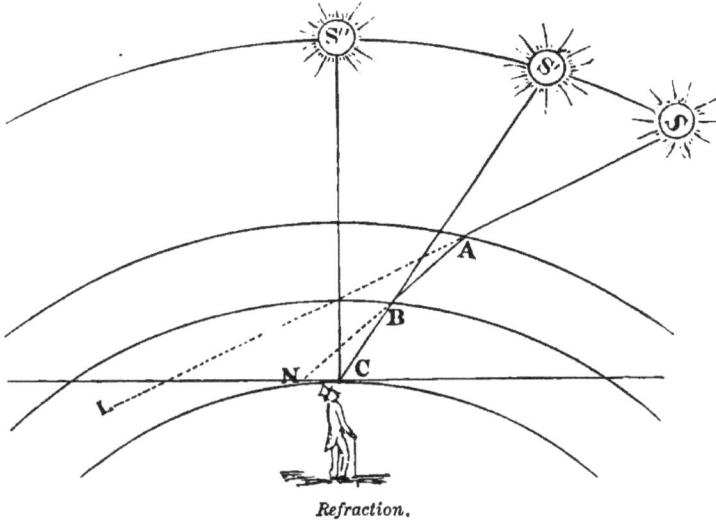

Refraction.

exceedingly rare. The rays of light from the heavenly bodies passing through these different layers are turned downward toward a perpendicular more and more as the density increases. According to a well-known law of optics (Physics, p. 150), if the ray of light from a star were bent in fifty directions before entering the eye, the star would nevertheless appear to be in the line of the one nearest the eye. The effect of this is, that the apparent place of a heavenly body is higher than the true place. The sun at S (Fig. 42), were it not for the atmosphere, would send a direct ray to L. Instead, the ray at A is refracted downward, and would then enter the eye at N ; passing, however, through a layer of a different density at B, it is again bent, and meets the eye of the observer at C. He sees the sun, not in the direction of the curved line C B A S, but in that of the straight line C B S.

The amount of refraction varies with the temperature, moisture, and other conditions of the atmosphere. It is zero for a body in the zenith, and increases gradually toward the horizon (as the thickness of the intervening atmosphere increases), where it is sometimes as much as 35'.

Fig. 43.

CHANGE OF PLACE AND APPEARANCE OF THE SUN AND THE MOON.—The sun may be really below the horizon, and yet seem to be above it. For example, on April 20, 1837, the moon was eclipsed before the sun had set. The mean diameter of both the sun and the moon being about half a degree, it follows that when we see the lower edge of either of these luminaries apparently just touching the horizon, in reality the whole disk is *below* it, and would be hidden were it not for the refraction. The day is consequently materially lengthened.

Fig. 44.

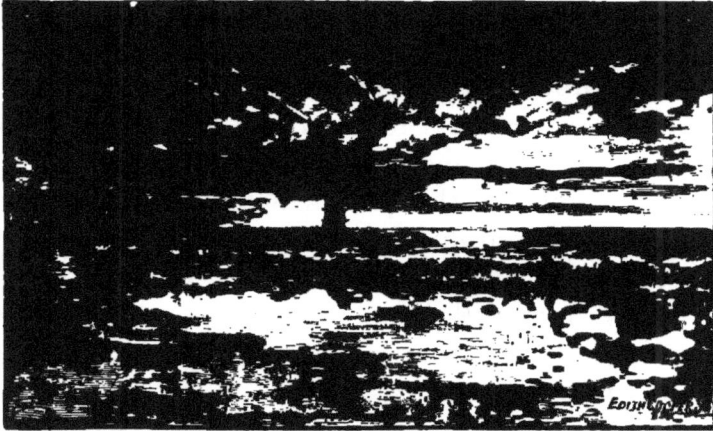

Deformation of the Sun near the Horizon.

The sun and the moon often appear *flattened* when near the horizon. The rays from the lower edge pass through a denser layer of the atmosphere, and are therefore refracted more than those from the upper edge: the effect of this is to make the vertical diameter appear less than the horizontal, and so to distort the figure of the disk into an oval shape.

The dim and hazy appearance of the heavenly bodies when near the horizon is caused not only by the rays of light having to pass a greater distance through the atmosphere, but also by their traversing the denser part. The intensity of the solar light is so greatly diminished by going through the lower strata, that we are then enabled to look upon the sun without being dazzled by his brilliant beams.

TWILIGHT.—The glow of light after sunset and before sunrise, which we term twilight, is caused by the refraction and the reflection of the sun's rays by the atmosphere. For a time after the sun has really set, the refracted rays continue to reach the earth; but when these have ceased, he still illuminates the clouds and upper strata of the air, just as he may be seen shining on the summits of lofty mountains long after he has disappeared from the view of the inhabitants of the plains below. The air and clouds thus illuminated reflect back a part of the light to the earth. As the sun sinks lower, less light reaches us, until reflection ceases and night ensues. The same thing occurs before sunrise, only in reverse order.

Twilight is usually reckoned to last until the depression of the sun below the horizon amounts to 18°; this, however, varies with the latitude,* seasons, and condition of the atmosphere. In the latitude of New York, twilight lasts from 1½ to 2 hours, the shortest twilight being in winter, and the longest in summer. Strictly speaking, in the latitude of Greenwich there is no true night for a month before and after the summer solstice, but constant twilight from sunset to sunrise. The sun is then near the Tropic of Cancer, and does not descend so much as 18° below the horizon during the entire night. At the equator the length of the evening twilight is about 1¼ hours, and remains almost con-

---

* When the sun's path is very oblique to the horizon, a longer time is required for the sun to descend or ascend the requisite vertical distance of 18° from the horizon; and a shorter time, when his path is more nearly perpendicular.

stant the entire year. The twilight is longest toward
the poles, where the night of six months is shortened
by an evening twilight of about fifty days and a
morning one of equal length.

DIFFUSED LIGHT.—The diffused light of day is produced in the same manner as that of twilight. The
atmosphere reflects and scatters the sunlight in
every direction. Were it not for this, no object
would be visible to us out of direct sunshine ; every
shadow of a passing cloud would be pitchy darkness ; the stars would be visible all day ; no window
would admit light except as the sun shone directly
through it, and a man would require a lantern to go
around his house at noon.

The blue light reflected to our eyes from the atmosphere above us, or, more correctly, from the
vapor in the air, produces the optical illusion we
call the sky. Were it not for this, every time we
cast our eyes upward we should feel like one gazing
over a dizzy precipice ; while now the crystal dome
of blue smiles down upon us so lovingly and beautifully that we call it heaven.

Aberration of Light.—We have seen that the
places of the heavenly bodies are apparently changed
by refraction. Besides this, there is another change
due to the motion of light combined with the motion
of the earth in its orbit. For example : the mean
distance of the earth from the sun is about 93,000,000
miles, and since light travels a little over 186,000
miles per second, it follows that the time occupied
by a ray of light in reaching us from the sun is
about 8¼ min. (8 min. 18 sec.) ; so that, in fact,

(1), we do not see the sun as it is, but as it was 8⅓ minutes ago. And since, during this time, the earth has moved in its orbit about 20½'' (2), we do not see that luminary in the exact place it occupies at the time of observation.

ILLUSTRATION.—Suppose a ball let fall from a point P, above the horizontal line A B, and a tube, of which A is the lower extremity, placed to receive it.

Fig. 45.

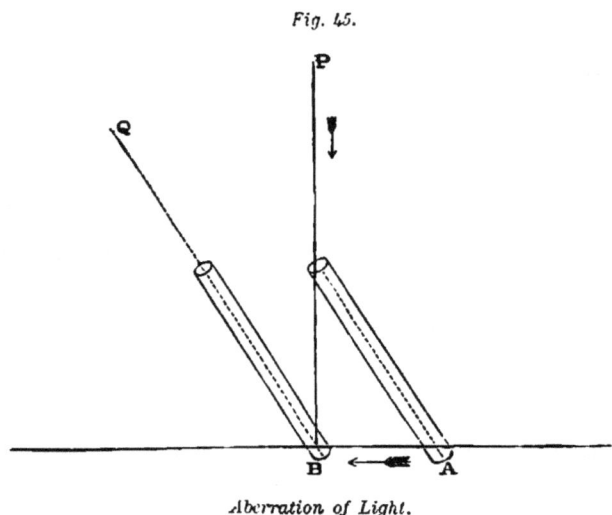

Aberration of Light.

If the tube were fixed, the ball would strike it on the lower side ; but if the tube were carried forward in the direction A B, with a velocity properly adjusted at every instant to that of the ball, while *preserving its inclination* to the horizon, so that when the ball, in its natural descent, reached B, the tube would have been carried into the position B Q, it is evident that the ball throughout its whole descent would be found in the tube ; and a spectator referring to the tube the motion of the ball, and carried along with the former, unconscious of its motion, would think that the ball had been moving in an inclined direction, and had come from Q.

A very common illustration may be seen almost any rainy day. Choose a time when the air is quiet, and the drops large. Then, if you stand still, you will see that the drops fall vertically ; but if you walk forward, you will see the drops fall as if they were meeting you. If, however, you walk backward, you will observe that the drops fall as if they were coming from behind you. We thus see that the drops have an apparent as well as a real motion.

THE GENERAL EFFECT OF ABERRATION is to cause each star apparently to describe in the course of a year a minute ellipse, the central point of which is the place the star would actually occupy were our globe at rest.

Parallax *is the difference in the direction of an object as seen from two different places.* For a simple illustration, hold your finger before you in front of the window. Upon looking at it with the left eye only, you will locate your finger at some point on the window ; on looking with the right eye only, you will locate it at an entirely different point. Use your eyes alternately and quickly, and you will be astonished to see how your finger will seem to change its place. Now, the difference in the direction of your finger as seen from the two eyes is its parallax.

In astronomical calculations, the position of a body as seen from the earth's surface is called its *apparent* place, while that in which it would be seen

from the center of the earth is called its *true*
place. Thus, in Fig. 46, a star is seen by the observer at O in the direction O P; if it could be
viewed from the center R, its direction would be
in the line R Q. It is therefore seen from O at a
point in the heavens *below* its position in reference
to R. From looking at the cut, we can see (1), that

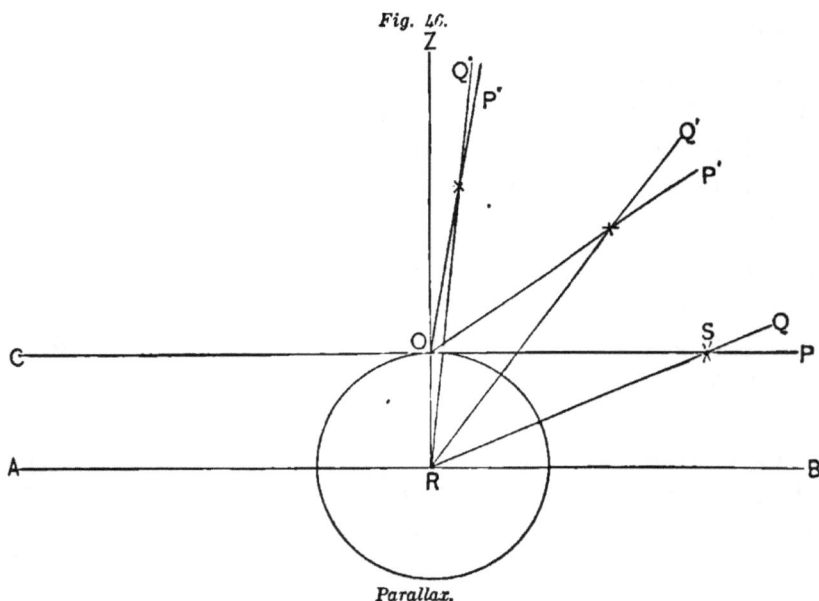

Fig. 46.

Parallax.

the parallax of a star near the horizon is greatest,
while it decreases gradually until it disappears altogether at the zenith, since an observer at O, as well
as one at R, would see the star Z directly overhead;
and (2), that the nearer a body is to the earth the
greater its parallax becomes.

It has been agreed by astronomers, for the sake of
uniformity, to correct all observations so as to refer

them to their true places as seen from the center of the earth. Tables of parallax are constructed for this purpose. The question of parallax is also of great importance, because as soon as the parallax of a body is accurately known, its distance, diameter, etc., can be determined. (See Celestial Measurements.)

HORIZONTAL PARALLAX is the parallax of a body when at the horizon. It is, in fact, *the earth's semi-diameter as seen from the body.* In Fig. 46, the parallax of the star S is the angle O S R, which is measured by the line O R—the semi-diameter of the earth. The *sun's horizontal parallax* is the angle subtended (measured) by the earth's semi-diameter as seen from that luminary. As the moon is nearest the earth, its horizontal parallax is greater than that of any other heavenly body.

ANNUAL PARALLAX.—The fixed stars are so distant from the earth that they exhibit no change of place when seen from different parts of the earth. The lines O S and R S are so long that they are apparently parallel. Astronomers, therefore, instead of taking the earth's semi-diameter, or 4,000 miles, as the measuring tape, observe the position of the fixed stars at opposite points in the earth's orbit. This gives a change in place of 186,000,000 miles. The variation of position which the stars undergo at these remote points is called their *annual parallax.*

6 .

THE MOON.

New Moon, ☽. First Quarter, ☾. Full Moon, ☉. Last Quarter, ●.

Motion in Space.—The orbit of the moon, considering the earth as fixed, is an ellipse of which our planet occupies one of the foci. Her distance from

Fig. 47.

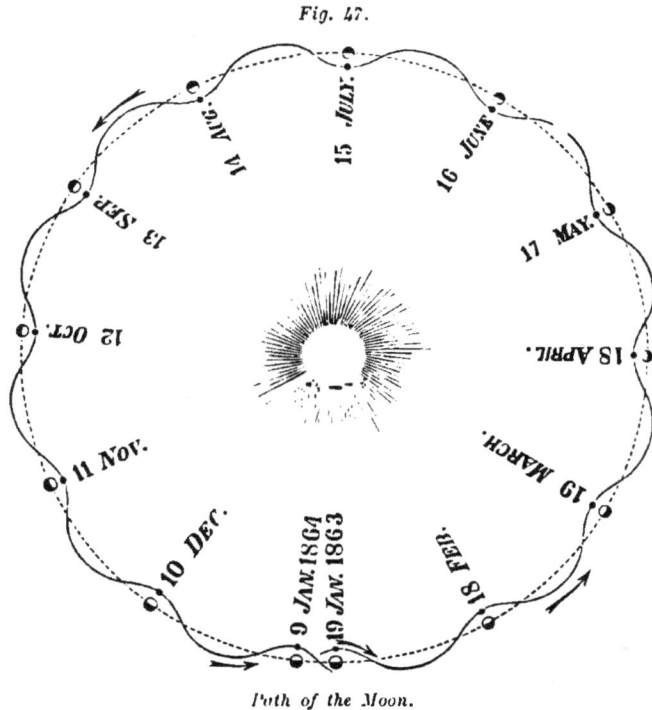

Path of the Moon.

the earth, therefore, varies incessantly. At perigee (*peri*, near ; *gē*, the earth), she is 26,000 miles nearer than in apogee (*apo*, from ; *gē*, the earth) : the mean distance is about 239,000 miles. To reach the moon, would require a chain of thirty globes equal in size to the earth. An ordinary express-train would take about a year to accomplish the journey.

The moon completes her revolution (*sidereal*) around the earth in about 27⅓ days; but, as the earth is constantly passing on in its orbit around the sun, it requires over two days longer before the moon comes into the same position with respect to the sun and the earth, thus completing a *synodic* revolution, or lunar month (29½ days).

THE REAL PATH OF THE MOON is the result of her own motion and the onward movement of the earth. The two combined produce a wave-like curve that crosses the earth's path twice each month; this, owing to its small diameter compared with that of the earth's orbit, is always concave toward the sun. As the moon constantly keeps the same side turned toward us, it follows that she must rotate on her axis once each month.

Dimensions.—The moon's diameter is about 2,160 miles. To equal the earth, would require fifty globes the size of the moon. The apparent size varies with the distance; the mean is, however, about one-half a degree, nearly the same as that of the sun. The moon always appears larger than she really is, on account of her brightness. This is the effect of what is termed in optics *Irradiation.** For the same reason it is often noticed that the crescent moon seems to be a part of a larger circle than the rest of

---

* To illustrate this principle, cut two circular pieces of the same size, one of black and the other of white paper. The white circle, when held in a bright light, will appear much larger than the black one.

the moon. The moon appears larger on the horizon than when high in the sky. This, however, is a mere illusion.* By an examination of the cut, it is easily seen that the moon is 4,000 miles nearer when on the zenith than when at the horizon.

Besides these general variations in size, the moon varies in apparent size to different observers. Much

Fig. 43.

The Distance of the Moon at the Horizon and at the Zenith.

amusement may be had in a large party or class by a comparison of her apparent magnitude. The estimates will differ from a small saucer to a wash-tub.

Librations (*librans,* swinging).—Though the moon presents the same hemisphere to us, there are three causes which enable us to see, in all, about $\frac{576}{1.000}$ of her entire surface.

1. The axis of the moon is inclined a little to her orbit, as also her orbit is inclined to the earth's orbit ; so, when her north pole leans alternately toward and

---

* At the horizon we compare her with various terrestrial objects which lie between her and us, while aloft we have no association to guide us in judging of her distance, and we are led to underrate her size. If we look at her when near the horizon, through a roll of paper, or the hands held tube-wise, this illusion will vanish.

from the earth, we see sometimes past her north and sometimes past her south pole. This is called *libration in latitude.*

2. The moon's rotation on her axis is always performed in the same time, while her movement along her orbit is variable; hence we occasionally see a little further around each *limb* (outer edge) than at other times. This is called *libration in longitude.*

3. The size of the earth is so much greater than that of the moon, that an observer, by the rotation of the earth, or by going north or south, can see further around the limbs.

Light and Heat.—If the whole sky were covered with full moons, they would scarcely make daylight, since the brilliancy of the moon does not exceed $\frac{1}{800,000}$ that of the sun. That portion of the moon's surface which is directly exposed to the sun has been thought to be highly heated, possibly to the degree of boiling water,* but this is now considered very improbable.

Whether or not the moon radiates any heat to the earth has long been a mooted question. The best authorities, at present, estimate the average heat of the moonbeam at about $\frac{1}{280,000}$ that we receive from the sun, or sufficient to raise the temperature of a sensitive black-bulb thermometer $\frac{1}{5,000}$ of a degree.

---

* Prof. Langley is now engaged in an exhaustive series of experiments upon this subject, using his famous "bolometer"—an instrument capable of detecting a difference of .00001° C. The result of his observations upon Mount Whitney (1881) showed that "mercury would remain a solid under the vertical rays of a tropical sun were radiation into space wholly unchecked, and that the temperature of a planet may, and not improbably does, depend far less upon its neighborhood to, or remoteness from, the sun, than upon the constitution of its atmosphere." As the moon has no air-blanket, it is therefore very doubtful whether its surface ever reaches a temperature of −100° F.

It would be absurd to suppose that this slight amount of heat can have any appreciable effect upon the weather.

Center of Gravity.—It is thought that possibly the center of gravity of the moon is not exactly at her center of magnitude, but about thirty-three miles beyond, the lighter half being toward us. If that be so, this side is equivalent to a mountain of that enormous height; and if water and air exist upon the moon, they cannot remain on this hemisphere, but must be confined to the side which is forever hidden from our view.

Atmosphere of the Moon.—The existence of an atmosphere upon our satellite is at present an open question. If there be any, it must be extremely rarefied, perhaps as much so as that in the vacuum obtained in the receiver of our best air-pumps.

Appearance of the Earth to Lunarians.—If there be any lunar inhabitants on the side toward us, the earth must present to them all the phases which their world exhibits to us, only in a reverse order. When we have a new moon, they have a *full earth*, a bright full-orbed moon fourteen times as large as ours. The lunar inhabitants upon the side opposite to us of course never see our earth, unless they take a journey to

Fig. 49.

Appearance of the Earth as seen from the Moon.

the regions from whence it is visible, to behold this wonderful spectacle. Those living near the limbs of the disk might, however, on account of the *libra-tions*, get occasional glimpses of it near their horizon.

The Earth-Shine.—For a few days before and after new moon, we may distinguish the outline of the unillumined portion of the moon. In England, it is popularly known as "the old moon in the new moon's arms." This reflection of the earth's rays must serve to keep the lunar nights quite light, even in *new earth*.

Phases of the Moon.—The phases of the moon show conclusively that it is a dark body, which shines by reflecting the light it receives from the sun. Let us compare its various appearances with the positions indicated in the figure.

(1.) We see the moon as a delicate crescent in the west just after sunset, as she emerges from the sun's rays at conjunction. She soon sets below the horizon. Half of the surface is illumined, but only a slender edge with the horns turned from the sun is visible to us. Each night the crescent broadens, the moon recedes about 13° further from the sun, and sets correspondingly later, until at quadrature half of the enlightened hemisphere is turned toward us, and the moon is said to be in her *first quarter*.

(2.) The moon, continuing her eastern progress round the earth, becomes *gibbous** in form, and, about the fifteenth day from new moon, reaches the point in the heavens directly opposite to that which

* *Gibbous* means more than a half and less than the whole of a circle.

the sun occupies. She is then in *opposition*, the whole of the illumined side is turned toward us, and we have a *full moon*. She is on the meridian at mid-

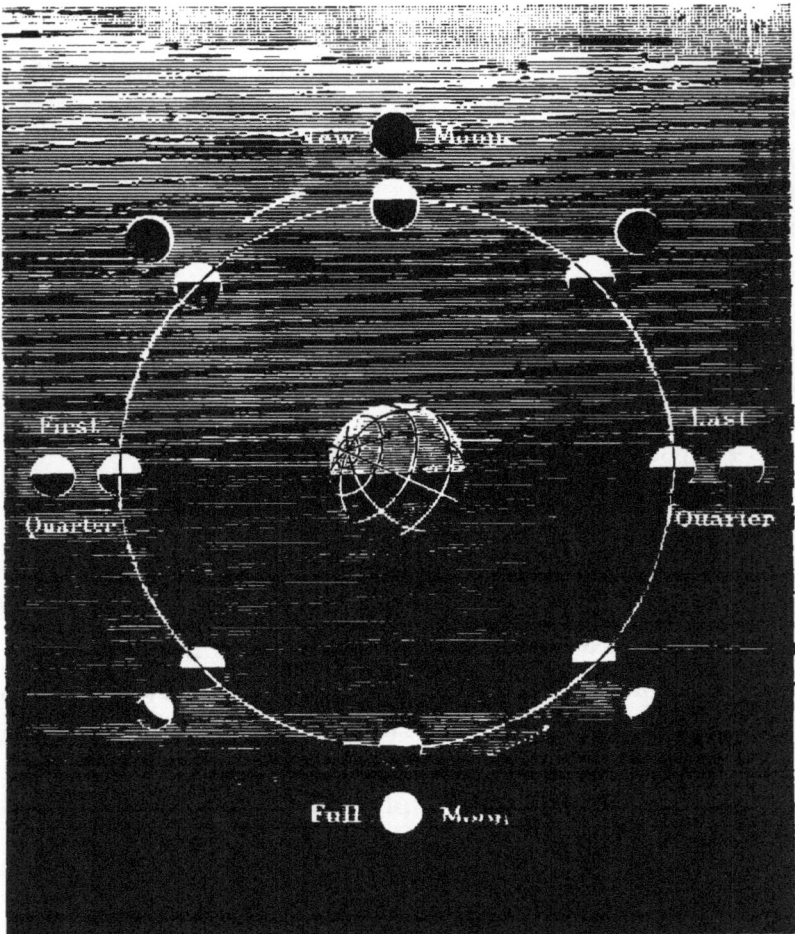

Phases of the Moon.

night, and so rises in the east as the sun sets in the west, and *vice versa*.

(3.) The moon, passing on in her orbit from opposition, presents phases reversed from those of the second quarter. The proportion of the illumined side visible to us gradually decreases ; she becomes *gibbous* again ; rises nearly an hour later each evening, and in the morning lingers high in the western sky after sunrise. She now comes into quadrature, and is in her *third quarter*.

(4). From the third quarter, the moon turns her enlightened side from us and decreases to the crescent form again ; as, however, the bright hemisphere constantly faces the sun, the horns are pointed toward the west. She is now seen as a bright crescent in the eastern sky just before sunrise. At last, the illumined side is completely turned from us, and the moon herself, coming into conjunction with the sun, is lost in his rays. To accomplish this journey through her orbit from new moon to new moon again, has required 29½ days—a *lunar month*.

MOON RUNS HIGH OR LOW.—All have, doubtless, noticed that, in the long nights of winter, the full moon is high in the heavens, and continues a long time above the horizon ; while in midsummer she is low, and remains a much shorter time above the horizon. This is a wise plan of the Creator, which is seen yet more clearly in the arctic regions. There, the moon, during the long summer day of six months, is above the horizon only her first and fourth quarters, when her light is least ; but during the tedious winter night of equal length, she is continually above the horizon for her second and third quarters. Thus, in polar regions, the moon is never full by day, but is always full every month in the night.

We can easily understand these phenomena when we remember that the new moon is in the same quarter with, and the full moon is in the opposite quarter from, the sun. When, therefore, the sun sinks low in the southern sky the full moon rises high, and when the sun rises high the full moon sinks low.

Harvest Moon.—While the moon rises, on the average, 50 m. later each night, the exact time varies from less than half an hour to a full hour. Near the time of the autumnal equinox the moon, at her full, rises about sunset for a number of nights in succession. This produces a series of brilliant moonlight evenings. It is the time of harvest in England, and hence has there received the name of the Harvest Moon. In the following month (October), the same occurence takes place; it is then termed the Hunter's Moon.

The cause of this phenomenon lies in the fact that the moon's path is variously inclined to the horizon at different seasons of the year.* When, at the time of rising, the full moon is near the vernal equinox, the angle her path makes with the horizon is least, and when she is near the autumnal equinox it is greatest. In the former case, the moon, moving eastward each day about 13°, will descend but little below the horizon, and so for several successive

---

* Besides this reason, we should remember that the motion of the moon is slowest at apogee and fastest at perigee. (See note, p. 302.)

evenings will rise at about the same hour. In the latter, she will descend much further each day and thus will rise much later each night. The least possible variation in the hour of rising is 17 minutes,— the greatest is 1 hour and 16 minutes.

In Figure 51, let S represent the sun ; E, the earth ; M, the moon ; C F, the moon's path around the earth when the autumnal equinox is in the eastern horizon ; E D, when the vernal equinox is in the eastern horizon ; A M B S, the horizon ; and M d = M b = 13°, the distance the moon moves each day. When passing along the path C F, the moon sinks below the horizon the distance $a\ b$, and when moving along the path E D, only the distance $c\ d$. It is obvious that before the moon can rise in the former case, the horizon

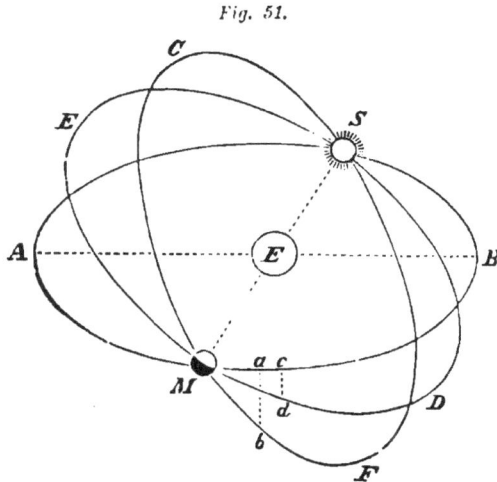

Fig. 51.

The Harvest Moon.

must be depressed the distance $a\ b$, and in the latter only $c\ d$; and the moon will rise each evening correspondingly later in the one and earlier in the other.

Cause of "Dry Moon," and "Wet Moon."—At new moon, when the bright crescent lies nearly perpendicular to the horizon, the moon is popularly called a *wet moon*, and when it is almost horizontal, the

moon is termed a *dry moon*. The cause of this
change in the crescent is astronomical, and not
meteorological. The form of the crescent has therefore no connection with the weather. A little reflection will show us that the horns, or cusps, of the
new moon must point from the sun. As the ecliptic
(from which the moon's path varies but slightly) is
differently inclined to the horizon at various times
of the year, this will give the crescent a different
position with reference to the horizon (p. 29).

Nodes.—The orbit of the moon is inclined to the
ecliptic about 5°, the points where her path crosses
it being termed *nodes*. The ascending node (☊) is
the place where the moon crosses in coming above
the ecliptic, or toward the north star; the descending
node (☋) is where it passes below the ecliptic. The
imaginary line connecting these two points is called
the "line of the nodes."

Occultation.—The moon, in the course of her
monthly journey round the earth, frequently passes
in front of the stars or planets, which disappear on
one side of her disk and reappear on the other. This
is termed an *occultation*, and is of practical use in
determining the difference of longitude between
various places on the earth.

Lunar Seasons; Day and Night, Etc.—As the
moon's axis is so nearly perpendicular to her orbit,
she cannot have any change of seasons. During
nearly fifteen of our days, the sun pours down his
rays unmitigated by any atmosphere to temper them.
To this long, torrid day succeeds a night of equal
length and polar cold.

Fig. 62.

Ideal Landscape on the Moon.

How strange the lunar appearance would be to us! The disk of the sun seems sharp and distinct. The sky is black and overspread with stars even at midday. There is no twilight, for the sun bursts instantly into day, and, after a fortnight's glare, as suddenly gives place to night; no air to conduct sound; no clouds; no winds; no rainbow; no blue sky; no gorgeous tinting of the heavens at sunrise and sunset; no delicate shading; no soft blending of colors, but only sharp outlines of sun and shade.*

The nights of the visible hemisphere must be brilliantly illuminated by the earth, whose phases "serve well as a clock—a dial all but fixed in the same part of the heavens, like an immense lamp, behind which the stars slowly defile along the black sky."

Telescopic Features.—Even with the naked eye, we see on the moon's surface bright spots (the summits of lofty mountains, gilded by the first rays of the sun), and darker portions—low plains yet lying in comparative shadow. The telescope reveals to us a region torn and shattered by fearful though now extinct† volcanic action. Everywhere the

---

* The moon is a fossil world, an ancient cinder, a ruined habitation perpetuated only to admonish the earth of her own impending fate, and to teach her occupants that another home must be provided, which frost and decay can never invade. The moon was once the seat of all the varied and intense activities that now characterize the surface of our earth. At one time its physical condition was like that of the parent earth from which it had just been separated: but, being smaller, it cooled faster, and its geologic periods were correspondingly shorter. Its life-age was perhaps reached while the earth was yet glowing.—Read Winchell's Geology of the Stars.

† Several distinguished astronomers assert, however, that the crater Linnæus has undergone noticeable transformations. Its sides seem to have fallen in, and the interior to have become filled up, as if by a new eruption. It is said to present an appearance similar to that of the Sea of Serenity. Other marked changes are said to have been discovered from time to time, on the moon's surface, but they are not generally ac-

Fig. 53.

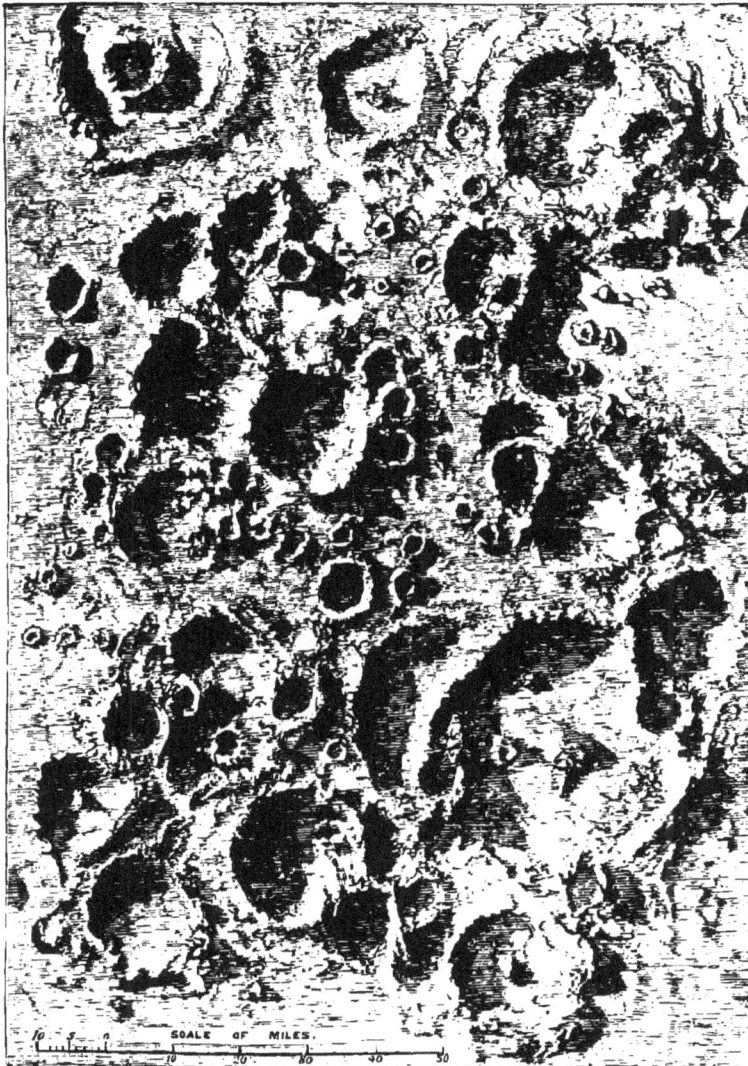

Telescopic View of the Moon.

crust is pierced by craters, whose irregular edges and rents testify to the convulsions our satellite has undergone.

Mountains.—The heights of more than 1,000 of the lunar mountains have been measured, some of which exceed 25,000 feet. When the sun's rays strike one of these mountains obliquely, the shadow is as distinctly perceived as that of an upright staff when placed opposite the sun. Some of the elevations are insulated peaks that shoot up from the center of circular plains ; others are mountain ranges extend-

Fig. 54.

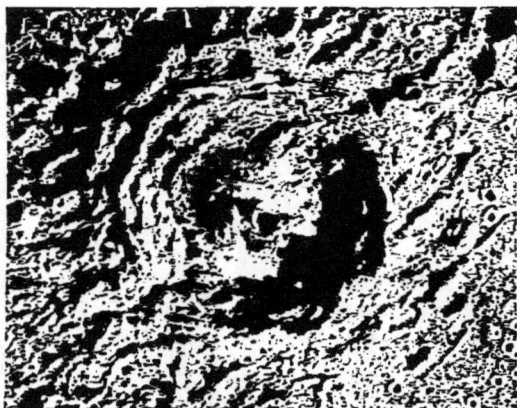

Copernicus.

ing hundreds of miles. Most of the lunar heights have received names of men distinguished in science. Thus we find Plato, Aristarchus, Copernicus,* Kepler.

credited. For an interesting discussion of this subject, read a chapter entitled "A New Crater in the Moon," in Proctor's Poetry of Astronomy.

    * This is one of the grandest of the lunar craters. It is situated on the tip of the nose of the "Man in the Moon." Its diameter is forty-six miles, and its encircling rampart rises 12,000 feet above the interior plateau, in the midst of which stands a group of cones. one 2,400 feet in height.

and Newton, associated, however, with the Apennines, Carpathians, etc.

GRAY PLAINS, OR SEAS.—These are analogous to our prairies. They were formerly supposed to be sheets of water, but they exhibit the uneven appearance of a plain, instead of the regular curve of a sea. The former names have been retained, and we find on lunar maps the Sea of Tranquillity, the Sea of Nectar, Sea of Serenity, etc.

RILLS, LUMINOUS BANDS. — The latter are long, bright streaks, irregular in outline and extent, which radiate in every direction from Tycho, Kepler, and other mountains; the former are similar, but are sunken, and have sloping sides, and were at first thought to be ancient river-beds. Their nature is a mystery.

CRATERS constitute the most curious feature of the lunar landscape. They are of volcanic origin, and usually consist of a cup-like basin, with a conical elevation in the center. Some of the craters have a diameter of over 100 miles, and are great walled plains, sunk so far behind huge, volcanic ramparts that the lofty wall surrounding an observer at the center would be beyond his horizon. Other craters are deep and narrow,—as Newton, which is said to be about four miles in depth,—so that neither earth nor sun is ever visible from a great part of the bottom. The appearance of these craters is strikingly shown in the accompanying view (Fig. 53) of the region to the southeast of Tycho.

ECLIPSES.

Eclipse of the Sun. — If the moon should pass through either node at or near the time of conjunction, or *new moon*, she would necessarily come between the earth and the sun, for the three bodies are then in the same straight line. This would cause

Fig. 55.

Theory of a Total and a Partial Eclipse of the Sun.

an eclipse of the sun. If the moon's orbit were in the same plane as the ecliptic, an eclipse of the sun would occur at every new moon ; but as the orbit is inclined, it can occur only at or near a node.

Fig. 56.

Theory of an Annular Eclipse of the Sun.

THE ECLIPSE MAY BE PARTIAL, TOTAL, OR ANNULAR. —In Fig. 55, we see where the dark shadow (*umbra*) of the moon falls on the earth and obscures the entire body of the sun. To the persons within that region, there is a *total eclipse ;* the breadth of this space is not large, averaging only 140 miles. Beyond this umbra, there is a lighter shadow, *penumbra,* (*pene,* almost; *umbra,* a shadow), where only a portion of the sun's disk is obscured. Within this region, there is a *partial eclipse.* To those persons living north of the equator and of the umbra, the eclipse passes over the lower limb of the sun ; to those south of the umbra, it passes over the upper limb.* When the eclipse occurs exactly at the node, it is said to be *central.* If the eclipse takes place when the moon is at apogee, her apparent diameter is less than that of the sun ; as a consequence, her disk does not cover the disk of the sun, and the visible portions of that luminary appear in the form of a ring (*annulus*) ; hence there is an *annular eclipse* in all those places comprised within the limits of the cone of shadow prolonged to the earth.

GENERAL FACTS CONCERNING A SOLAR ECLIPSE.—The following data may guide in understanding the phenomena of solar eclipses.

(1.) The moon must be new.

(2.) She must be at or near a node.

(3.) When her distance from the earth is less than the length of her shadow, the eclipse will be total or partial.

(4.) When her distance is greater than the length of her shadow, the eclipse will be annular or partial.

---

* South of the equator the reverse of these phenomena would happen.

(5.) There can be no eclipse at those places where the sun himself is invisible.

(6.) An eclipse is not visible over the whole illumined side of the earth. As the moon's diameter is less than that of the earth, her cone of shadow is too small to enshroud the entire globe, so that the region in which it is total cannot exceed 180 miles in breadth. As, however, the earth is constantly rotating on its axis during the duration of the eclipse, the shadow may travel over a large surface.

(7.) If the moon's shadow fall upon the earth when she is nearing her ascending node, it will sweep

Fig. 57.

Solar Ecliptic Limit (17°).

across the south polar regions : if, when nearing her descending node, it will graze the earth near the north pole. The nearer a node a conjunction occurs, the nearer the equatorial regions the shadow will strike.

(8.) At the equator, the longest possible duration of a total solar eclipse is about eight minutes ; of an annular, twelve minutes. One reason of the greater length of the latter is, that then the moon is in apogee, when she always moves slower than in perigee. The duration of total obscuration is greatest when the moon is in perigee and the sun in apogee ; for then the apparent size of the moon is greatest, and that of the sun is least.

(9.) There cannot be more than five nor less than two solar eclipses per year. A total or an annular eclipse, in its recurrence at any place, is exceedingly rare. There has been (according to Halley) only one total eclipse visible at London since 1140.

(10.) A solar eclipse comes on the western limb, or edge of the sun, and passes off on the eastern.

(11.) The disk of the sun is divided into twelve digits, and the amount of the eclipse is estimated by the number of digits which it covers. Thus an eclipse of six digits is one in which half the diameter of the disk is concealed.

CURIOUS PHENOMENA attend a total eclipse. Around the sun is seen a beautiful corona, or halo of light, like that which painters give to the head of the Virgin Mary. Flames of a rose-red color play around the disk of the moon. When only a mere crescent of the sun is visible, it seems to resolve itself into bright spots interspersed with

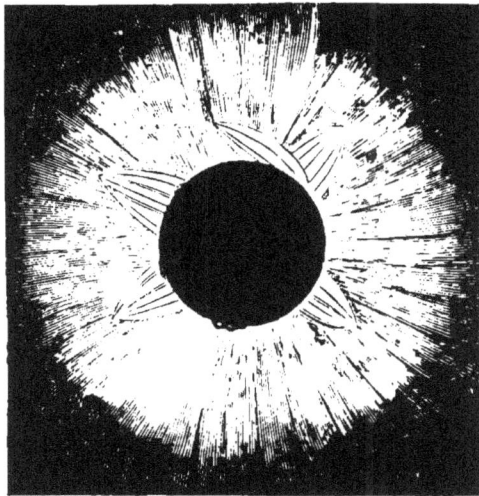

Fig. 58.

Eclipse of 1858.

dark spaces, having the appearance of a string of glittering beads (Baily's Beads).

The attendant circumstances of a total eclipse

are of a peculiarly impressive character. The darkness is so dense that the brighter stars and planets are seen, birds cease their songs and fly to their nests, flowers close, and the face of nature assumes an unearthly, cadaverous hue, while a sudden fall of the temperature causes the air to feel damp, and the grass to be wet as if from excessive dew. Orange, yellow, and copper tints give objects a strange appearance. "Men look at each other, and behold, as it were, corpses."

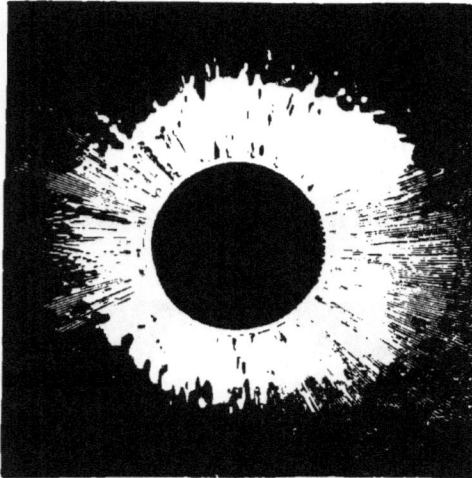

Fig. 59.

Annular Eclipse of 1835, showing Baily's Beads.

The ancients regarded a total eclipse with feelings of indescribable terror, as an indication of the anger of an offended Deity, or the presage of some impending calamity.* Even now, when the causes

* William of Malmesbury thus connects the eclipse of August 2, 1133, with Henry I., who left England on that day, never to return alive: "The elements manifested their sorrows at this great man's last departure. For the sun on that day, at the 6th hour, shrouded his glorious face, as the poets say, in hideous darkness, agitating the hearts of men by an eclipse: and on the 6th day of the week, early in the morning, there was so great an earthquake that the ground appeared suddenly to sink down; an horrid noise being first heard beneath the surface."

The same writer, speaking of the total eclipse of March 20, 1140, says: "During this year, in Lent, on the 13th of the kalends of April, at the 9th hour of the 4th day of the week, there was an eclipse, throughout England, as I have heard. With us, indeed, and with all our neighbours, the obscuration of the sun also was so remarkable, that persons sitting at table, as it then happened almost every where, for it was Lent, at first feared

Fig. 60.

Corona seen in 1871.

are fully understood, and the time of the eclipse can
be predicted within the fraction of a second, the

that Chaos was come again : afterwards learning the cause, they went out and beheld
the stars around the sun. It was thought and said by many, not untruly, that the king
(Stephen) would not continue a year in the government."

Columbus made use of an approaching eclipse of the moon, which took place March 1,
1504, to relieve his fleet, then in great distress from want of supplies. As a punishment
to the islanders of Jamaica, who refused to assist him, he threatened to deprive them
of the light of the moon. At first they were indifferent to his threats, but " when the
eclipse actually commenced, the barbarians vied with each other in the production of
the necessary supplies for the Spanish fleet."

Among the Hindoos a singular custom is said to exist. When, during a solar eclipse,
the black disk of our satellite begins slowly to advance over the sun, the natives believe
that some terrific monster is gradually devouring it. Thereupon they beat gongs, and
rend the air with screams of terror and shouts of vengeance. For a time their frantic

change from broad daylight to almost instantaneous gloom is overwhelming, and inspires with awe even the most careless observer. (See note, p. 303.)

The Saros.—The nodes of the moon's orbit are constantly moving backward. They complete a revolution around the ecliptic in about 18¼ years. Now the moon makes 223 synodic revolutions in 18 years and 10 days ; the sun makes 19 revolutions with regard to the lunar nodes in about the same time. Hence, in that period, the sun, the moon, and the nodes will be in nearly the same relative position. If, then, we reckon 18 years and 10 days from any eclipse, we shall find the time of its repetition.

This method was discovered, it is said, by the Chaldeans. The ancients were enabled, by this means, to predict eclipses, but it is considered too inaccurate by modern astronomers.

Metonic Cycle.—The Metonic Cycle (sometimes confounded with the Saros) was not used for foretelling eclipses, but for ascertaining the *age of the moon* at a given period. It consists of nineteen tropical years,* during which time there are 235 new moons ; so that, at the end of this period, the new moons will recur at seasons of the year corresponding to those of the preceding cycle. By registering, therefore, the exact days of any cycle at which the

efforts seem futile and the eclipse still progresses. At length, however, the increasing uproar reaches the voracious monster ; he appears to pause, and then, like a fish rejecting a nearly swallowed bait, gradually disgorges the fiery mouthful. When the sun is quite clear of the great dragon's mouth, a shout of joy is raised, and the poor natives disperse, delighted to think that they have so successfully relieved their deity from his impending peril.

* A tropical year is the interval between two successive returns of the sun to the vernal equinox.

PLATE III.

Various Forms of Solar Prominences. (*See* pp. 53, 141, 262.)

new and full moons occur, such a calendar shows on what days these events will happen in succeeding cycles.

Since the appointment of games, feasts, and fasts has been made very extensively, both in ancient and modern times, according to new or full moons, such a calendar becomes very convenient for finding the day on which the required new or full moon takes place. Thus, if a festival were decreed to be held in any given year on the day of the first full moon after the vernal equinox : find what year it is of the lunar cycle, then refer to the corresponding year of the preceding cycle, and the day will be the same. The *Golden Number*, a term still used in our almanacs, denotes the year of the lunar cycle. Four is the golden number for 1884.

Fig. 62.

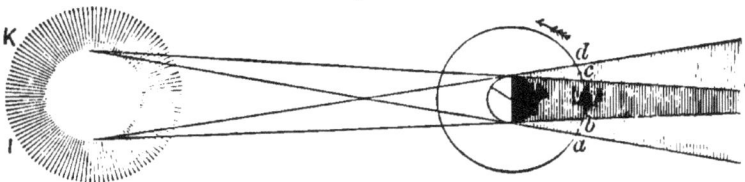

Eclipse of the Moon.

An Eclipse of the Moon is caused by the passing of the moon into the shadow of the earth, and hence can take place only at full moon—*opposition.* As the moon's orbit is inclined to the ecliptic, her path is partly above and partly below the earth's shadow ; thus an eclipse of the moon can take place only at or near one of the nodes. In Fig. 62, the *umbra* is represented by the space between the lines K c and

7

I *b* ; outside of this is the *penumbra*, where the earth cuts off the light of only a portion of the sun. The moon enters the *penumbra* of the earth at *a*,—this is termed her *first contact with the penumbra ;* next she encounters the dark shadow of the earth at *b*,—this is called the *first contact with the umbra ;* she then emerges from the umbra at *c*,—the *second contact with the umbra ;* finally, she touches the outer edge of the penumbra at *d*,—the *second contact with the penumbra.* Since the earth is so much larger than the moon, the eclipse can never be *annular ;* as, however, the eclipse may occur a little above or below the node, the moon may only partly enter the earth's shadow, either on its upper or lower limb. From the first to the last contact with the penumbra, five hours and a half may elapse.

Total eclipses of the moon are rarer events than those of the sun, since the lunar ecliptic limit is only about 12° ; yet they are more frequently seen by us, (1) because each one is visible over the entire unillumined hemisphere of the earth, and also (2) because by the diurnal rotation during the long duration of the eclipse, large areas may be brought within its limits. So it will happen that while the inhabitants of one district witness the eclipse throughout its continuance, those of other regions merely see its beginning, and others only its termination.

The moon does not completely disappear even in total eclipses. The cause of this lies in the refraction of the solar rays in traversing the lower strata of the earth's atmosphere ; they are analyzed, and purple our moon with the tints of sunset. The amount of refraction and the color depend upon the state of the air at the time.

THE TIDES.

Description.—Twice a day, at intervals of about twelve hours and twenty-five minutes, the water begins to set in from the ocean, beating the pebbles and the foot of the rocky shore, and dashing its spray high in air. For about six hours, it climbs far up on the beach, flooding the low lands and transforming creeks into rivers. The instant of *high-water* or *flood-tide* being reached, the water begins to descend, and the *ebb* succeeds the *flow*. The water, however, falls somewhat slower than it rises.

Fig. 63.

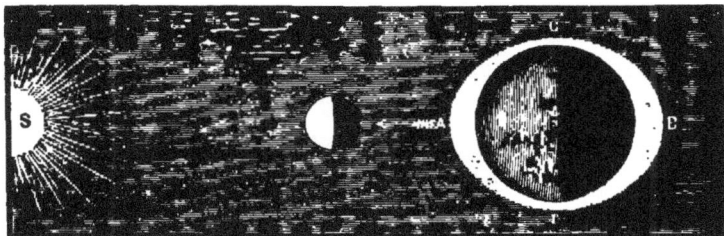

Spring Tide.

The Tides are Caused by a great wave, which, raised by the moon's attraction, follows her in her course around the earth.* The sun, also, aids some-

* Prof. Ball, Royal Astronomer of Ireland, claims that once the moon was nearer the earth than now; the day and the month were equal, each three hours long. At 40,000 miles distance, the moon was a greater tide producer by 216 times. As the moon receded from the earth, both revolved more slowly. At the present time, 27 earth rotations equal one moon rotation. This has remained, and will remain, sensibly true, for thousands of years. But the friction of the tides will, in the far future, lengthen the day to equal 57 of our present days,—a condition that will then last for ages.

what in producing this effect; but as the moon is 400
times nearer the earth, her influence is far greater.*

As the waters are free to yield to the attraction of
the moon, she draws them away from C and D and
they become heaped up at A. The earth, being
nearer the moon than the waters on the opposite
side, is more strongly attracted, and so, being drawn
away from them, they are left heaped up at B. As
the result, high-water is produced at A by the water
being pulled from the earth, and at B by the earth
being pulled from the water.

The influence of the moon requires a little time to
produce its full effect; hence high-water does not
occur at any place when the moon is on the meridian, but a few hours after. As the moon rises
about fifty minutes later each day, there is a corresponding difference in the time of high-water.
While, however, the lunar tide-wave thus lags about
fifty minutes every day, the solar tide occurs uniformly at the same time. They therefore steadily
separate from each other. At one time, they coincide, and high-water is the sum of the lunar and
solar tides; at other times, high-water of the solar
tide and low-water of the lunar tide occur simultaneously, and high-water is then the difference between the lunar and solar tides.†

* The whole attraction of the moon is only $\frac{1}{13}$ that of the sun: yet her influence in
producing the tides and precession is greater, because that depends not upon the *entire*
attraction either exerts, but upon the *difference* between their attraction upon the
earth's center and upon the earth's nearest surface. For the moon, on account of her
nearness, the proportion of the distance of these parts is treble that of the sun, and
hence her greater effect.

† We should bear in mind the philosophical truth, that the tide in the open sea does
not consist of a progressive movement of the water itself, but only of the form of the
wave.—*Physics, p.* 101,

CAUSES THAT MODIFY THE TIDES.—At new and at full moon (the *syzygies*) the sun acts with the moon (Fig. 63) in elevating the waters ; this produces the highest, or *Spring-tide*. In quadrature (Fig. 64), the sun tends to diminish the height of the water : this is called *Neap-tide*. When the moon is in perigee, her attraction is stronger ; hence the flood-tide is higher, and the ebb-tide is lower than at other times.

Fig. 64.

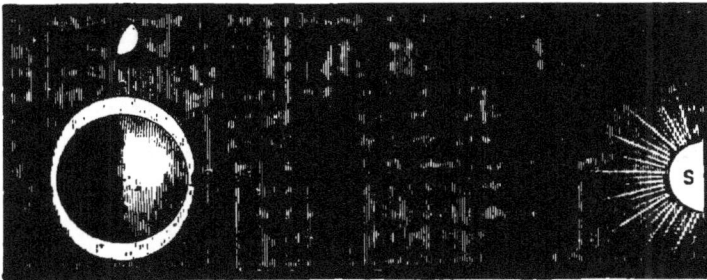

Neap Tide.

This remark applies also to the sun. The height of the tide also varies with the declination of the sun and the moon,—the highest or equinoctial tides taking place at the equinoxes, if, when the sun is over the equator, the moon also happens to be very near it : the lowest occur at the solstices. The force and the direction of the winds, the shape of the coast, and the depth of the sea greatly complicate the explanation of local tides.

HEIGHT OF THE TIDE AT DIFFERENT PLACES.—In the open sea, the tide is hardly noticeable, the water sometimes rising not higher than a foot ; but where the wave breaks on the shore, or is forced up into

bays or narrow channels, it is very conspicuous.
The difference between ebb and flood neap-tide at
New York is over three feet, and that of spring-tide
over five feet; while at Boston it is nearly double
this amount. A headland jutting out into the ocean
will diminish the tide; as, for instance, off Cape
Florida, where the average height is only one and a
half feet. A deep bay opening up into the land like
a funnel will converge the wave, as at the Bay
of Fundy, where it rolls in, a great, roaring wall
of water sixty feet high, frequently overtaking and
sweeping off men and animals.* The tide sets up
against the current of rivers, and often entirely
changes their character; for example, the Avon
at Bristol is a shallow ditch, but at flood-tide it
becomes a deep channel navigable by the largest
Indiamen.

V. MARS.

The god of war. Sign, ♂, shield and spear.

Description.—Passing outward in our survey of the
solar system, we next meet with Mars. This is the
first of the *superior* planets, and the one most like
the earth. It appears to the naked eye as a bright
red star, rarely scintillating, and shining with a
steady light, which distinguishes it from the fixed

* The tide-wave ascends the Hudson River at about the same speed as the steamboats; at Albany it reaches a height of a little over two feet.

stars.* At conjunction its apparent diameter is only
about 4″; but once in about two years it comes into
opposition with the sun, when its diameter may increase to 30″. At intervals of nearly 15 years, this
occurs when the planet is in perihelion and the earth
in aphelion. Mars then shines with a brilliancy
rivalling that of Jupiter himself.†

Fig. 65.

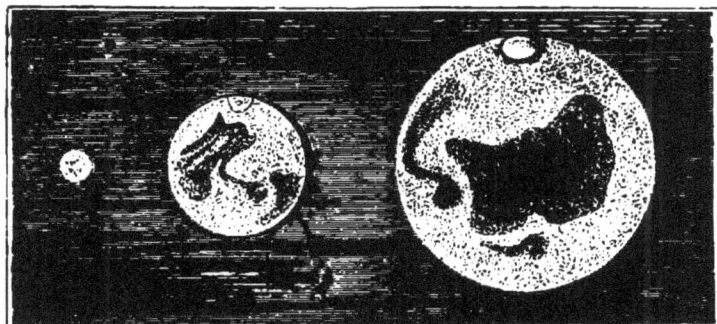

Diameter of Mars at Extreme, Mean, and Least Distances.

Motion in Space.—Mars revolves around the Sun
at a mean distance of about 141,000,000 miles. Its
orbit is sufficiently flattened to bring it at perihelion
26,000,000 miles nearer that luminary than when in
aphelion. Its motion varies in different portions of
its orbit, but the average velocity is about fifteen
miles per second. The Martian day is 37 min. longer
than ours, and the year contains about 668 Martian
days, equal to 687 terrestrial days (nearly two years).
Distance from Earth.—When in opposition, the

* Its ruddy appearance has led to its being celebrated among all nations. The Jews
gave it the appellation of "blazing," and it bore in other languages a similar name.

† The next favorable opposition will occur in 1892.

distance of Mars is (like that of all the superior planets) the difference between the distance of the planet and that of the earth from the Sun : at conjunction, it is the sum of these distances. If the orbits were circular, these distances would be the same at every revolution. The elliptical figure, however, occasions much variation. Thus, if Mars, at opposition, be in perihelion while the earth is in aphelion, it is removed from us about 34,000,000 miles.

Dimensions.—The diameter of Mars is nearly 4,200 miles.* Its volume is about $\frac{1}{7}$ and its density $\frac{4}{5}$ that of the earth. A stone let fall on its surface would fall six feet the first second. It is somewhat flattened at the poles, and bulged at the equator, like our globe.

Seasons.—The light and heat of the sun at Mars are less than one-half that which we enjoy. Its axis is inclined about 27°, therefore its zones and seasons do not differ materially from our own : its days, also, as we have seen, are of nearly the same length. Since, however, its year is equal to nearly two of our years, the seasons are lengthened in proportion.

There must be a considerable difference between the temperature of its northern and southern hemispheres, as the former has its summer when 26,000,000 miles further from the sun than the latter : an increased length of 76 days may, however, be sufficient

* Some authors place the *diameter of Mars at about 5,000 miles.* There is, also, a discrepancy as to the other data of this planet. Prof. Hall, as the result of his observations, gives the density = .776 ; force of gravity = .37 ; fall of a body, 1st sec. = 6 feet. With the discovery of the satellites we have now the means of securing exact results. The difficulty of observation, however, is shown from the fact that " the light which falls upon the earth from one of these moons is about what a man's hand on which the sun shines at Washington would reflect to Boston."

compensation. It has an atmosphere like our own, loaded with clouds.

Mars has two moons.* Our earth and its moon present in the Martian evening sky a beautiful pair of planets, constantly remaining in close proximity to each other, and exhibiting all the phases which Mercury and Venus present to us.

Telescopic Features.—Under the telescope, Mars exhibits slight phases. Its surface is covered with reddish spots, which are believed to be continents.† Other portions, of a greenish tint, are considered to

Fig. 66.

View of Mars.

be bodies of water. The proportion of land to water on the earth is reversed in Mars. "*Here* every continent is an island; *there* every sea is a lake: but

---

* The satellites of Mars were discovered in August, 1877, by Prof. Hall of the Naval Observatory, Washington. The outer one revolves about the planet in 30 hr. 18 min., at a distance of about 12,300 miles; and the inner one in 7 hr. 40 min., at a distance of 3,600 miles (less than that of remote cities on our own continent). The inner moon moves so much faster than the rotation of Mars that to an inhabitant of that planet, the moon would seem to rise in the west and set in the east, passing through all the phases of our moon during a single night. The moons have been named Deimos and Phobus, or Dread and Terror—the sons of Mars. The diameter of these little globes is probably less than 15 miles. For an amusing description of such a world, read "Living in Dread and Terror," a chapter in Proctor's "Poetry of Astronomy."

† So carefully has the surface of this planet been studied, that a globe of Mars has been prepared which is said to be in some respects more perfect than any globe of the earth. The different bodies of land and water have been named after distinguished astronomers. A characteristic feature of the seas is the long, narrow channels. Schiaparelli, the Italian astronomer, claims to have discovered a number of singular dark lines, now known as "canals." They seem to connect different bodies of water, and, though without sufficient reason, have been by some considered as the work of the Martian inhabitants.

these, like our own continents, are chiefly confined to one hemisphere, so that the habitable area of the two globes may not differ so much as the size of the planets."

The ruddy color is thought by Herschel to be due to an ochery tinge in the soil; by others it is attributed to peculiarities of the atmosphere and clouds. Lambert suggests that the color of the vegetation on Mars may be red instead of green. There are constant changes going on in the brightness of the disk, owing, it is supposed, to the variation of the clouds of vapor in its atmosphere. No mountains have yet been discovered.

In the vicinity of the poles are brilliant white spots, which are considered to be masses of snow. The *"snow zones"* apparently melt and recede with the return of summer in each hemisphere, and increase on the approach of winter. We can thus from the earth watch the formation of polar ice and the fall of snow,—in fact, the changes of the seasons—on the surface of a neighboring planet.

VI. THE MINOR PLANETS.

Discovery.—Beyond Mars there is a wide interval that was not filled until the present century. The bold, imaginative Kepler conjectured that there was a planet in this space. This supposition was corroborated by Titius's discovery of what has since been known as

Bode's Law.—Take the numbers 0, 3, 6, 12, 24, 48,

96, 192, 384, each of which, after the second, is double the preceding one. If we add 4 to each of these numbers, we form a new series :

4, 7, 10, 16, 28, 52, 100, 196, 388.

At the time this law was discovered, these numbers represented very nearly the proportionate distance from the sun of the planets then known, taking the earth's distance as ten, except that there was a blank opposite 28. This naturally led to inquiry, and a systematic effort to solve the mystery.*

On the 1st day of January, 1801, the nineteenth century was inaugurated by Piazzi's discovery of the small planet Ceres, at almost the exact distance necessary to fill the gap in Bode's series. The announcement of other new planets soon followed, until now (1885) there are two hundred and forty-seven, with a probability of more being found. Indeed, Leverrier has calculated that there may be perhaps 150,000 in all.

Description. — These minor worlds, or "pocket planets," as Herschel styled them, are diminutive indeed. The largest of them is Vesta, which shines at times as a star of the 6th magnitude, and can then be seen with the naked eye.† Those recently discov-

---

* It is a curious fact that the discovery of Ceres should have been made by an outsider, as Piazzi did not belong to the society of 24 astronomers then searching for the planet. The publication of Bode's law had little to do with the result. In fact, the direct cause was an error of the press in putting an extra star in Wollaston's Catalogue, and while Piazzi was looking for this star he found Ceres.

† The small size of the disks of the minor planets defies exact measurement. Newcomb makes Ceres and Vesta the largest of the group, with diameters between 200 and 400 miles. Echo has been assigned a diameter of 17 miles, or not far from the size of the miniature moons of Mars. Several of these little worlds have been found but to be lost again ; while the mere labor of tracing the movements of so many tiny globes already surpasses the probable worth of the results.

ered are so small that it is difficult to decide which is the smallest. A good walker could easily make the tour of one in a day ; a prairie farmer would need to pre-empt a whole such world for a cornfield. "A man placed on one of these tiny globes could leap C0 feet high, and, in his descent, would sustain no greater shock than he does on the earth from jumping or leaping a yard." These planets revolve around the sun in regular orbits, comprising a zone about 100,000,000 miles in width. Their paths are variously inclined to the ecliptic ; Massalia's is only 41', while that of Pallas rises 34°.

Origin.—A conjecture concerning the origin of these bodies is, that they are the fragments of a large planet that, in a remote antiquity, was shivered to pieces by some terrible catastrophe. "One fact seems above all others to confirm the idea of an intimate relation between these planets. It is this : if their orbits consisted of solid rings, they would be found so entangled that it would be possible, by taking up any one at random, to lift all the rest." The more probable view is given under the "Nebular Hypothesis."

NAMES AND SIGNS.—Ceres, the first discovered, received the symbol ⚳, a sickle, as that goddess was supposed to preside over harvests. Pallas, the second, named from the goddess of wisdom and scientific warfare, obtained the sign ⚴, the head of a spear. Of late, a simple circle with the number inclosed has been adopted ; thus ① represents Ceres, ② is the sign of Pallas.

VII. JUPITER.

The king of the gods.   Sign ♃, a hieroglyphic representation of an eagle, "the bird of Jove."

Description.—From the smallest members of the solar system we now pass to the largest planet—the colossal Jupiter. Its peculiar splendor and brilliancy distinguish it from the fixed stars, and vie even with the lustre of Venus. It is one of the five planets discovered in primitive ages.*

Motion in Space.—Jupiter revolves about the sun at a mean distance of about 483,000,000 miles. His movement among the fixed stars is slow and majestic, comporting well with his vast dimensions and the dignity conferred by four attendant worlds. He advances through the zodiac at the rate of one sign yearly; so that if we locate the planet now, a year hence we shall find it equally advanced in the next sign. Yet slowly as he seems to travel through the heavens, he is bowling along through space at the enormous speed of nearly 500 miles per minute. The Jovian day is equal to only about ten of our hours, while the year is lengthened to about 12 of our years, comprising near 10,000 of his days.

Distance from Earth.—Once in thirteen months Jupiter is in *opposition,* and his distance from the earth is measured by the difference of the distances of the two bodies from the sun. At the expiration

* In those early times, Jupiter was supposed to be the cause of storm and tempest. Pliny thought that lightning owed its origin to this planet. An old almanac of 1508, foretelling the harmless condition of Jupiter for a certain month, says, "Jubit es hote and moyste and does weel til al thynges and noyes nothing."

of half this time he is in *conjunction*, and his distance from us is measured by the sum of these distances.

Dimensions.—The diameter of this planet is about 90,000 miles. Its volume is 1,400 times that of the earth, and much exceeds that of all the other planets combined. Seen at the distance of the moon, this immense globe would embrace 1,000 times the space of the full moon. Its density is only one-quarter that of the earth : moreover, its rapid rotation upon its axis, whereby a particle on

Fig. 67.

View of Jupiter.

the equator revolves with a velocity of 473 miles per minute against the earth's 17 miles per minute, must produce a powerful centrifugal force which materially diminishes the weight of objects near its equator. Consequently, a stone let fall on Jupiter would pass through only about 42 feet the first second. As a result of this rapid rotation, the planet is one of the most flattened of any in the solar system, the equatorial diameter exceeding the polar by 5,000 miles.

Seasons.—As the axis of Jupiter is but slightly inclined from a perpendicular to the plane of its orbit, there is little difference in the length of his days and nights, which are each of about five-hours duration. At the poles, the sun is visible for nearly six years, and then remains set for the same length of time. The seasons are but slightly varied. Summer reigns near the equator, while the temperate regions enjoy perpetual spring. The light and heat of the sun are only $\frac{1}{27}$ of what we receive; yet peculiarities of soil or atmosphere may compensate this difference. The evening sky on Jupiter must be magnificent; besides the glittering stars which adorn our heavens, four moons, waxing and waning, each with its diverse phase, illuminate his night. All the starry exhibition sweeps through the sky in five hours.

Telescopic Features.—JUPITER'S MOONS.—Through the telescope* Jupiter presents a beautiful Copernican system in miniature. Four small stars—moons—accompany him in his twelve-yearly revolutions. From hour to hour their positions vary, and they seem to oscillate from one side to the other of the planet. At one time, there will be two on each side; and again, three on one side, while the remaining star is left alone. They are also frequently found to disappear, one, two, or even three at a time, and, more rarely, all four at once.

These moons are called by the ordinal numbers, reckoning outward from the planet. With an ordinary glass, there is nothing to distinguish them from small stars. The IIIrd., being the largest and

---

* There are well-authenticated instances on record of their having been seen by the naked eye. Among others, the following singular case is mentioned. Wrangle, the celebrated Russian traveler, states that, when in Siberia, he once met a hunter, who said, pointing to Jupiter, "I have just seen that star swallow a small one and then vomit it up again."

brightest, will generally be identified the most easily.
The Ist. satellite appears to the inhabitants of the
planet almost as large as our moon to us; the IInd.,
and IIIrd., about half as large.

SATELLITES OF JUPITER.

	Mean distance from Jupiter.	Diameter.	Density. Water as 1.	Sidereal period.		
				D.	H.	M.
I. Io..............	267,380	2,352 m.	1.12	1	18	23
II. Europa..........	425,156	2,009 "	2.14	3	13	4
III. Ganymede........	678,393	3,436 "	1.87	7	3	43
IV. Callisto..........	1,192,823	2,929 "	1.47	19	16	32

It is noticeable that here are four satellites revolving about Jupiter, one of them larger than the
planet Mercury, and each surpassing in size the
minor planets between Mars and Jupiter. The
moons are not only distinguised by their various
dimensions, but also by the variety of their color.
The Ist. and IInd. have a bluish tint, the IIIrd. a
yellow, and the IVth. a reddish shade. The space
occupied by this miniature system is about two and
a half million miles in diameter.

ECLIPSE OF THE MOONS.—Jupiter, like all celestial
bodies not self-luminous, casts into space a cone of
shade. The Ist., IInd., and IIIrd. satellites revolve
in orbits but very little inclined to the plane of the
planet's orbit. During each revolution, they pass
between the Sun and Jupiter, producing a solar
eclipse; and also, by passing through the shadow of
the planet itself, cause to themselves an eclipse
of the sun, and to Jupiter an eclipse of a moon.
The IVth. moon passes through a path more inclined, and therefore its eclipses are less frequent;

instead of being fully eclipsed, it sometimes just grazes the shadow. Through a telescope, we can distinctly watch the disappearance, or *immersion*, of the

Fig. 68.

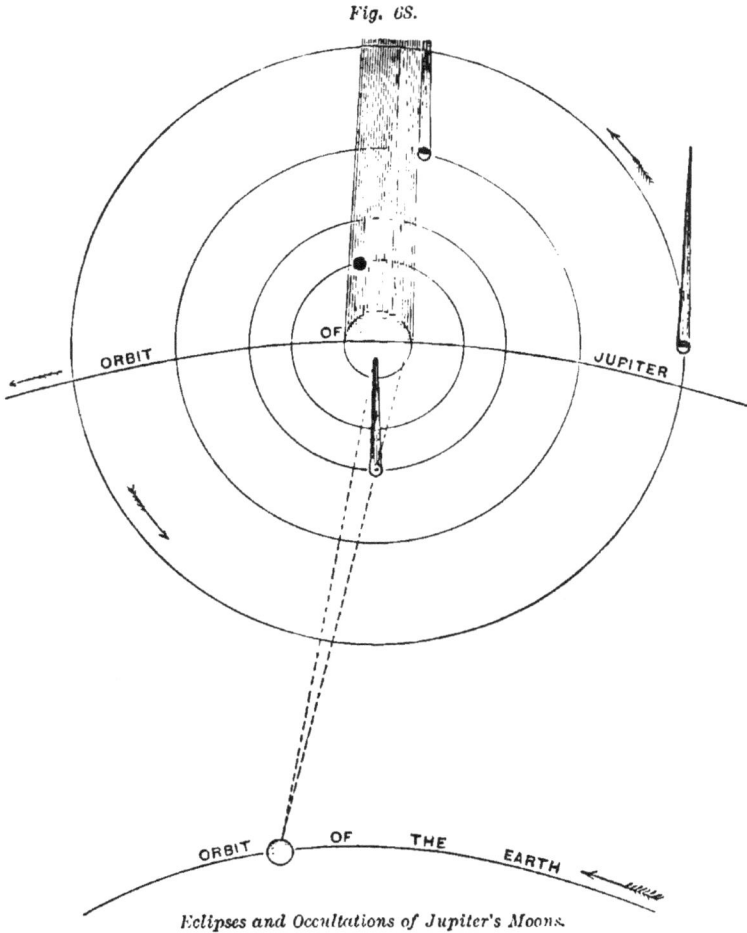

Eclipses and Occultations of Jupiter's Moons.

satellites in the planet's shadow, their reappearance, or *emersion*, and also the *transits* of their shadows as round black dots moving across the disk of Jupiter.

In Fig. 68, we see various positions of the moons : the Ist. is eclipsed ; the IInd. is passing across the disk of the planet on which its shadow is also thrown ; the IIIrd. is just behind the planet, and so *occulted* or concealed, while it has not yet entered the shadow; the IVth. is in view from the earth.

These satellites revolve with great rapidity, as is necessary in order to overcome the superior attraction of the planet and prevent their being drawn to its surface. The Ist. goes through all its phases in $1\frac{3}{4}$ days ; the IVth., in less than twenty days. A spectator on Jupiter might witness, during the Jovian year, 4,500 eclipses of the moon (moons), and about the same number of the sun.

Velocity of Light.—By an attentive examination of the eclipses of Jupiter's moons, Römer (a Danish astronomer), in 1617, discovered that the motion of

Fig 69.

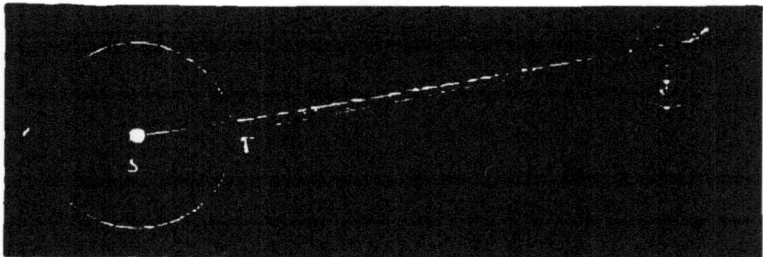

light is not instantaneous, as was then believed. He noticed that the observed times of the eclipses were sometimes earlier and sometimes later than the calculated times, according as Jupiter was nearest or furthest from the earth. In Fig. 69, let J represent Jupiter ; *e*, one of the moons ; S, the sun ; and T and *t*, different positions of the earth in its orbit. When the earth is at T, the eclipse occurs 16 min. and 36 sec. earlier than at *t*. That interval of time is required for the light to travel across the earth's orbit, giving a velocity of about 186,000 miles per second.

JUPITER'S BELTS are dusky streaks of varying breadth and number, lying more or less parallel to the planet's equator. A brighter, often rose-colored, space marks the equatorial regions. The belts are not permanent, but change sometimes in the course of a few hours. Occasionally, only two or three broad belts are visible ; at other times, a dozen narrow ones appear. Often, spots are seen that are more lasting than the dark stripes.* It is now supposed that the planet is enveloped in dense clouds, through which light cannot penetrate, and that the globe itself is heated to a high degree, and gives off vapors.† The parallel appearance is doubtless due to strong equatorial currents, analogous to our trade-winds.

* In 1878, a "Great Red Spot" appeared in the southern hemisphere of Jupiter. Its length was estimated at 8,000 miles, and its breadth at 2,000 miles. This curious phenomenon is still visible, but much diminished in brightness (1884).

† Jupiter and Saturn are older planets than the earth and Mars, but, being so large, they have cooled more slowly, and are yet only partially solidified, so that Jupiter, at least, still shines with much of its primeval fire. Mars typifies the middle age ; Saturn and Jupiter, the youth ; and Uranus and Neptune, the infancy of planetary existence. In the case of Saturn and Jupiter, we never see the real planets, but only the outline of their atmospheres. If this theory be true, Jupiter and Saturn now represent the condition in which our earth existed ages ago, before a solid crust had been formed upon its surface.—(*Geology*, p. 17.)

VIII. SATURN.

The god of time. Sign ♄, an ancient scythe.

Description.—We now reach, in our outward journey from the sun, the most remote world known to
the ancients. It shines with a steady pale yellow
light, which distinguishes it from the fixed stars.
Its orbit is so vast that its movement among the
constellations may be easily traced through one's
lifetime. It requires two and a half years to pass
through a single sign of the zodiac ;* hence, when
once known, it may be readily found again. The
earth leaves it at conjunction, makes a yearly revolution about the sun. comes to its starting point,
and overtakes Saturn in about thirteen days thereafter.† It is smaller than Jupiter, but more gor-
geously attended. Besides a retinue of eight satellites, it is surrounded by a system of rings, some
shining with a golden light, and others transparent,
—a spectacle as wonderful as it is unique.

Motion in Space.—Saturn revolves about the sun
at a mean distance of nearly 886,000,000 miles. The
eccentricity of its orbit is a trifle more than that of
Jupiter, so that while it may, at perihelion, come
fifty million miles nearer than its mean distance, at
aphelion it swings off as much beyond. We can
form some estimate of the size of its immense orbit,
when we remember that it is moving 22,000 miles

---

* Because of its slow, dreary pace, Saturn was chosen by the ancients as the symbol
for lead.

† From this the year of Saturn may be determined. As 13 : 378 days : : Earth's
year : Saturn's year = 30 yr. nearly.

per hour, and yet, from night to night, we can
scarcely detect any
change of place.
The Saturnian year
is equal to about
thirty of ours, and
comprises nearly
25,000 Saturnian
days, each about
10¼ hours long.

Fig. 70.

Saturn.

The Distance from the Earth is found in the same
manner as that of the other superior planets, being
least in opposition and greatest in conjunction. According as the earth and Saturn occupy different
portions of their orbits, the distances between them
at different times may vary nearly 300,000,000 miles.

Dimensions. — The diameter of Saturn is about
73,000 miles. Its volume is 700 times that of the
earth. Its density is about ¾ that of water, or a
little more than that of pine wood. The Saturnian
force of gravity is therefore scarcely greater than
the terrestrial, so that a stone would fall toward
the surface of that immense globe only about seventeen feet the first second.

Seasons.—The light and heat of the sun at Saturn
are only $.\frac{1}{100}$ that which we receive. The axis of
Saturn is inclined from a perpendicular to the plane
of its orbit about 31°.* The seasons therefore are

* Proctor says 26° ; others, 28°. Let the pupil adapt the paragraph to each of these
estimates.

similar to those of the earth, but on a larger scale. The sun climbs in summer about 8° higher above the horizon, and sinks correspondingly lower in winter. The tropics are 16° further apart, and the arctic and antarctic circles 8° further from the poles. Each of Saturn's seasons lasts more than seven of our years. There is an interval of fifteen years between the autumn and spring equinoxes, and between the summer and winter solstices. For fifteen years, the sun shines on the north pole, and a night of the same length envelopes the south pole. The atmosphere is doubtless very dense, as the belts seem to indicate.

Telescopic Features. — SATURN'S RINGS. — Galileo first noticed something peculiar in the shape of Saturn. Through his imperfect telescope it seemed to have on each side a small planet, like a supporter, to help old Saturn on his way. Galileo therefore announced to his friend Kepler the curious discovery, that "Saturn is threefold." As the planet, however, approached its equinoxes, these attendants vanished from his instrument. This was a great perplexity to the philosopher, and he never solved the mystery. When the rings were afterward seen, their real form was not known. They were supposed to be a kind of *handle* attached to the planet.

Description of the Rings.—The series consists of three rings of unequal breadth, surrounding the planet at the equator. The exterior ring is separated from the middle one by a distinct break, while the interior ring seems joined to the middle one. They differ in their brightness; the exterior ring is of a grayish tint; the middle one is the most brilliant, being more luminous than Saturn itself ; the interior one is darker and has a purple tinge. The two outer rings are known as the *bright* rings, and the inner one is called the *dusky* ring. The exterior and middle rings are both opaque and cast on the planet a distinct shadow ; while the interior one is so transparent that it appears upon the globe of Saturn as a dark band through which the surface of the planet is readily seen.

SATURN'S RINGS. (Proctor.)

	Miles.
Diameter of exterior ring	166,920
Breadth of exterior ring	10,000
Diameter of middle ring	144,300
Breadth of middle ring	17,600
Distance between exterior and middle ring	1,700
Diameter of interior ring	92,000
Breadth of interior ring	8,600
Distance of interior ring from the planet	10,000
Entire breadth of ring system	37,570
Thickness of rings, less than	100

Rotation.—The rings revolve around Saturn in about 10½ hours, in the same direction as the planet rotates on its axis. The globe of Saturn is not exactly at the center of the rings. This fact, combined with the rotary motion, is essential to the stability of the rings, preventing them from being precipitated upon the planet.

Phases of the Rings.—The plane of the rings is inclined about 28° to the ecliptic. In its revolution about the sun, the axis of Saturn remaining parallel to itself, the sun sometimes illumines the northern and sometimes the southern face of the rings. At Saturn's equinoxes, only the edge receives the light,

and the rings are invisible to us, except with the
most powerful telescopes, and then only as a line of
light. The body of the planet constantly cuts off
the sun's rays from a portion of the rings, and also
serves to conceal from our view some of the lumin-

Fig. 71.

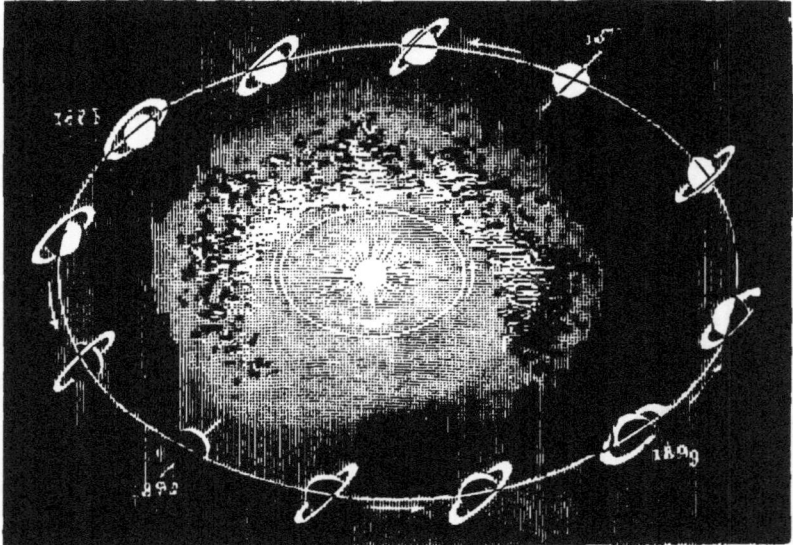

Phases of Saturn's Rings.

ous part. By a careful study of the cut, these various
positions of the planet and rings, with the favorable
times for observation, may be understood.

Composition of the Rings.—It is now generally
believed that the rings consist of a cloud of tiny
satellites,—too small to be seen with the telescope,—
revolving about the planet (see Nebular Hypothesis).

BELTS.—The surface of Saturn is traversed by faint

dusky belts of a far less distinct and definite appearance than those upon Jupiter. The equatorial regions are more strongly marked than the other parts of the disk.

COMPOSITION OF THE PLANET.—It is quite probable that Saturn, like Jupiter, has no solid crust, but consists of molten matter surrounded by vapor that continually rises from the heated interior (note, p. 163).

Satellites.—Saturn has eight satellites.

Number.	Names of Saturn's Satellites.	Distance from Saturn in miles.	Approximate diameter in miles.	Sidereal Period in days.
I.	Mimas.	120,800	1,000	0.94
II.	Enceladus.	155,015	?	1.37
III.	Tethys.	191,248	500	1.88
IV.	Dione.	245,876	500	2.73
V.	Rhea.	343,414	1,200	4.51
VI.	Titan.	796,157	3,300	15.94
VII.	Hyperion.	1,006,656	?	21.29
VIII.	Japetus.	2,313,835	1,800	79.33

Titan is the largest, and in size exceeds Mercury. Enceladus and Mimas are the faintest of twinklers, and can be seen only with a powerful telescope. They were first detected by Herschel, "threading like pearls the silver line of light," to which the ring, then seen edgewise, was reduced,—advancing off it at either end, returning, and then hiding themselves behind the planet. The first three of these moons are nearer to Saturn than our moon is to the earth, but Japetus is nearly ten times as distant: so that the diameter of the Saturnian system is nearly four and a half million miles.

Saturnian Scenery. — The magnificence of the

scenery upon Saturn must surpass anything with
which we are familiar.   In the cut, is given an ideal
view of a landscape located upon the planet at a latitude of about 28°, taken at midnight.   The rings form
an immense arch, which spans the sky and sheds a

Fig. 72.

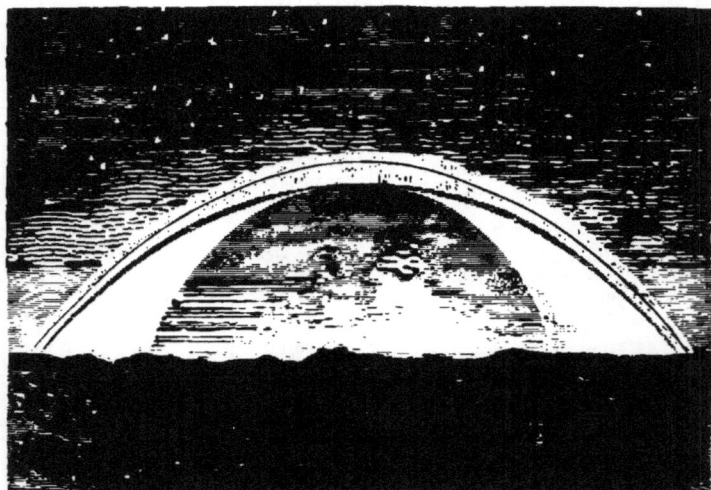

Ideal Landscape on Saturn, supposing a solid crust to exist.

soft radiance around ; while, to add to the strange
beauty of the night, eight moons in all their different
phases—full. new, crescent, or gibbous—light up the
starry vault.

IX. URANUS.

" Heaven," the most ancient of the gods. Sign, ♅ ; H, the initial letter of Herschel,
with a planet suspended from the cross-bar.   ;

Description.—On the 13th of March, 1781, between
10 and 11 P. M., Sir William Herschel was examining

with his great telescope some stars in the constellation Gemini. A small star attracting his attention, he observed it with a higher magnifying power, when, unlike the fixed stars, its disk widened. Watching it for several nights, he detected its motion in space; but, mistaking its true character, he announced the discovery of a new comet. A few-months examination revealed the error, and the new body—new to us, but older perhaps than our own world*—was admitted to be a member of the solar system.

Uranus may be seen in a dark sky, by a person of strong eyesight, if he previously knows its exact position among the stars. Its faintness is due to its great distance from the earth. Were it as near as the sun, it would appear twice as large as Jupiter.

Motion in Space.—Uranus revolves about the sun at a mean distance of nearly 1,782,000,000 miles. Its year exceeds eighty-four of ours.

Dimensions.—Its diameter is about 33,000 miles. Its density is about equal to that of the water from the Dead Sea. The force of gravity upon the surface of the planet is $\frac{9}{10}$ that upon the earth.

Seasons.—We know little of the seasons of Uranus. If its axis lies in the plane of its orbit, the sun must wind in a spiral form around the planet. The light and heat are less than $\frac{3}{1,000}$ of that which we

---

* It is now known that Uranus had been previously observed by other astronomers. Le Monier at Paris had watched it for twelve successive nights, but pronounced it a fixed star. He had also seen it on previous occasions, and had he been an orderly observer, he would doubtless have detected its planetary character; but he was extremely careless, as may be inferred from the fact related by Arago, that he had been shown one of Le Monier's observations of this planet written on a paper bag which originally contained hair-powder purchased at a perfumer's.

receive; the light has been estimated to be about the
quantity that would be afforded by three hundred
full moons.   The inhabitants of Uranus, if any such
exist, can see Saturn, and perhaps Jupiter, but none
of the planets within the orbit of the latter.

Telescopic Features.—No spots or belts have been
discovered.   The time of rotation and the other
features so familiar to us in the nearer planets are
therefore unknown with regard to Uranus.

Satellites.—Uranus has four moons, of which
little is known except the curious fact that their
orbits are nearly perpendicular to the plane of the
planet's orbit, and that their movements are apparently retrograde—*i. e.*, in the same direction as the
hands of a watch.

X. NEPTUNE.

The god of the sea.   Sign, ♆ , his trident.

Description.—Neptune is the far-off sentinel at the
outpost of the solar system, being the most distant
planet of which we have any knowledge.   It is invisible to the naked eye, and appears in the telescope
as a star of the sixth magnitude.

Discovery.—For many years, the motions of Uranus had been such as to baffle the most perfect calculations.   While far-distant Saturn, after his journey of
thirty years, came around to his place true to the minute, Uranus defied arithmetic, and refused to con-
form to the time set down for him on the heavenly
dial.

At length it was suggested that there was another planet exterior to Uranus, whose attraction produced these perturbations. So marked was this impression with Herschel, that he writes: "We see it as Columbus saw America from the shores of Spain. Its movements have been felt trembling along the far-reaching line of our analysis with a certainty not far inferior to ocular demonstration."

Finally, two young mathematicians, Leverrier, of Paris, and Adams, of Cambridge, England, each unknown to the other, set about the task of finding the place of this new planet. The problem was this: *Given the disturbances produced by the attraction of the unknown planet, to find its orbit and its place in the orbit.*

Adams, after assiduous labor for nearly two years, completed his calculations and submitted them to Prof. Airy, the Astronomer Royal, in 1845. In the summer of 1846, Leverrier laid a paper before the Academy of Sciences in Paris, announcing the position of the unknown planet. Prof. Airy, hearing of this, was so impressed with the value of Adams's calculations, that he wrote to Prof. Challis, of Cambridge, to search that quarter of the heavens. Prof. Challis did as requested, and saw a star which afterward proved to be the planet so anxiously sought for, although at that time he failed to ascertain its true character. In September, of the same year, Leverrier wrote to Berlin, asking for assistance in searching for the planet. Dr. Galle, on receiving the request, turned the large telescope of the Observatory to the place indicated, and almost immediately detected a bright star not laid down in the maps. This proved to be the predicted planet, found within less than a degree of the spot described by Leverrier.

Such is the history of one of the grandest achievements of the human mind. It stands as an ever fresh and assuring proof of the exactness of astronomical calculations, and the power of the intellect to understand the laws of the God of Nature.

Motion in Space.—Neptune revolves about the sun at a mean distance of about 2,790,000,000 of miles. The Neptunian year is equal to nearly 165 terrestrial ones. Its motion in its orbit is the slowest of any of the planets, since it is the most remote from the sun. The velocity decreases from Mercury, which moves at the rate of about 105,000 miles per hour, to Neptune, whose rate is only 12,000 miles.

Dimensions.—Neptune's diameter is about 37,000 miles. Its volume is nearly 100 times that of the earth. Its density is a little less than that of Uranus.

Seasons.—As the inclination of its axis is unknown, nothing can be ascertained concerning its seasons. The sun gives to Neptune but $\frac{1}{1,000}$ the light and heat which we receive.

Though Neptune is at the extreme of the solar system, 2,790,000,000 miles beyond us, the same heavens bend above, the Milky Way is no nearer to the eye, and the fixed stars shine no more brightly. The planets, however, are all too near the sun to be seen, except Saturn and Uranus. The Neptunian astronomers, if there be any, are well situated for measuring the annual parallax of the stars, since

Neptune has an orbit of 5,580,000,000 miles in diameter, and hence the angle must be thirty times as great as that which the terrestrial orbit affords.

Telescopic Features.—On account of the recent discovery of this planet and its immense distance, nothing is known of its rotation or physical features.

SATELLITES.—Neptune has one moon, at nearly the same distance from it as our own moon is from the earth. The revolution of this body about the planet, which is accomplished in about six days, has furnished the materials for calculating the mass of Neptune.

III. METEORS AND SHOOTING STARS.

Description.—All are familiar with those luminous bodies that flash through our atmosphere as if the stars were indeed falling from heaven. Different names have been applied to them, although the distinction is not very definite.

(1) AEROLITES are those stony or iron masses which descend to the earth.

(2) METEORS are luminous bodies which have a sensible diameter and a spherical form. They frequently pass over a great extent of country, and are seen for some seconds. Many leave behind them a train of glowing sparks ; others explode with reports like the discharge of artillery,—the pieces either continuing their course, or falling to the earth as

aërolites. Some meteors pass on into space; some are vaporized; while others are burned, and the ashes and fragments fall to the ground.

(3) SHOOTING STARS are those evanescent, brilliant

Fig. 73.

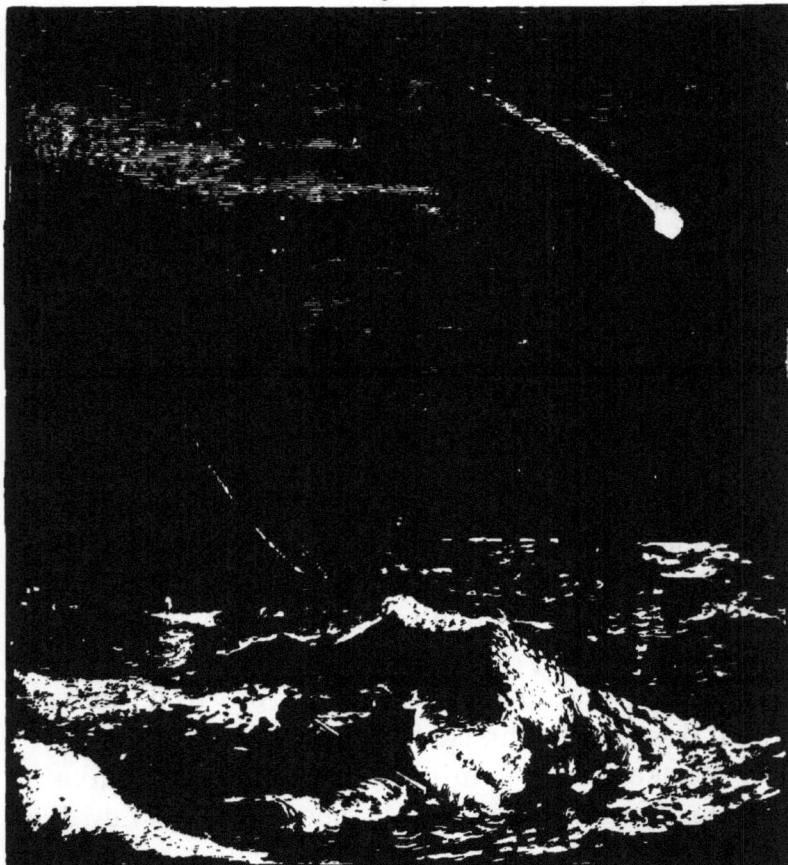

A Meteor with its Train.

points that suddenly dart through the higher regions of the air, leaving a fiery train behind.

1. Aerolites.—The fall of aërolites is frequently mentioned and well authenticated. Chinese records tell of one as long ago as 616 B.C., that, in its fall, broke several chariots and killed ten men. A block of stone, equal to a full wagon-load, fell in the Helles-pont, B.C. 465. By the ancients, these stones were held in great repute. The Emperor Jehangire, it is related, had a sword forged from a mass of meteoric iron which fell in the Punjab in 1620. In 1795, a mass was seen, by a ploughman, to descend not far from where he was standing. It threw up the soil on every side, and penetrated some distance into the solid rock beneath. In 1807, there was a shower of stones, one weighing 200 lbs., at Weston, Connect-icut. A mass once fell in South America, that was estimated to weigh fifteen tons. When first dis-covered, it was so hot as to prevent all approach. Upon its cooling, many efforts were made, by some travelers who were present, to detach specimens, but its hardness was too great for the tools that they possessed. In Yale College cabinet, there is a mass of meteoric iron, weighing 1,635 lbs.

AEROLITES CONSIST OF ELEMENTS which are famil-iar. The analysis of these stellar objects gives us names as commonplace as if they had known a far less romantic origin,—iron, tin, copper, nickel, cobalt, lime, magnesia, oxygen, sulphur, phos-phorus; in all, about twenty elements have been found. This fact is interesting as revealing some-thing of the chemistry of the region of space, concern-ing which we otherwise know little. The compounds, however, are so peculiar as to distinguish an aërolite from other substances. For example, meteoric iron, a prominent constituent of aërolites, is an alloy that has never been found in terrestrial minerals.

2. Meteors.—The records of meteors are even more wonderful than those of aërolites.   It is related that at Crema, Italy, one day in the 15th century, the sky at noonday became dark,—a cloud of appalling blackness overspreading the heavens.   Upon this cloud, appeared the semblance of a great peacock of fire

Fig. 74.

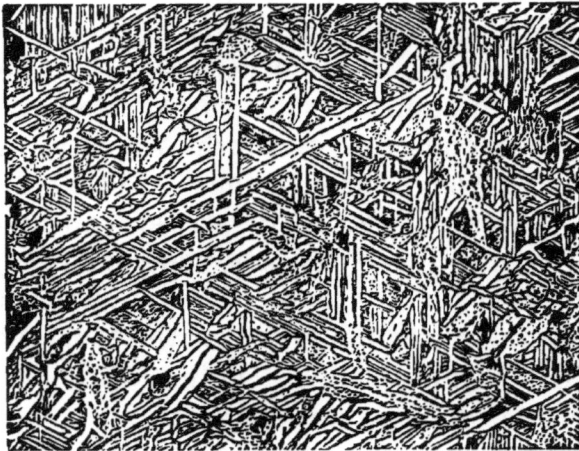

Copy of a Print Showing the Peculiar Crystalline Structure of Meteoric Iron.

flying over the town.   This suddenly changed to a huge pyramid, that rapidly traversed the sky. Thence arose awful lightnings and thunderings, amid which there fell upon the plain rocks, some of which weighed 100 lbs.   In 1803, a brilliant fireball was seen traversing Normandy with great velocity, and some moments after, frightful explosions, like the noise of cannon, were heard coming from a black cloud hanging in the clear sky; they were prolonged for five or six minutes. These discharges were followed by a shower of heated stones, some weighing over 24 lbs. In 1819, a meteor was witnessed in Massachusetts and Maryland, the diameter of which was estimated at half a mile. In July, 1860, a brilliant fireball passed over the State of New York, from west to east, and was last seen far out at sea. On the evening of Feb. 12th, 1875, a magnificent meteor "illumined the entire State of Iowa, and parts of Missouri, Illinois, Wisconsin, and Minnesota. The aërolites that have been collected show its weight to have been fully 5,000 lbs."

3. **Shooting Stars.**—One of the earliest accounts of star-showers is that which relates how, in 472, the sky at Constantinople appeared to be alive with flying stars and meteors. In some Eastern annals we are told that in October, 1202, "the stars appeared like waves upon the sky. They flew about like grasshoppers, and were dispersed from left to right." It is recorded that in the time of King William II. there occurred in England a wonderful shower of stars, which "seemed to fall like rain from heaven. An eye-witness, seeing where an aërolite fell, cast water upon it, which was raised in steam, with a great noise of boiling." *

SHOWERS OF 1799 AND 1833.—The most remarkable accounts are those of the showers of November 12th, 1799, and November 13th, 1833. Humboldt, in describing the former, says the sky was covered

* Rastel says concerning it : "By the report of the common people in this kynge's time, diverse great wonders were seene, and therefore the kynge was told by diverse of his familiars that God was not content with his lyvyng."

with innumerable fiery trails, which incessantly traversed the sky. From the beginning of the phenomenon, there was not a space in the heavens three times the diameter of the moon that was not filled every instant with the celestial fireworks,—large meteors blending constantly their dazzling brilliancy with the long phosphorescent paths of the shooting stars. (See notes, p. 305.)

The latter shower was most brilliant on this continent, and was visible from the lakes to the equator. Phosphoric lines swept over the sky like the flakes of a snow-storm. Large meteors darted across the heavens, leaving luminous trains behind them that were visible sometimes for half an hour : they generally shed a soft white light ; occasionally, however, yellow, green, and other colors varied the scene. Irregular fireballs. almost stationary, glared in the sky ; one especially, larger than the moon. hung in mid-air over Niagara Falls, and mingled its light with the foam and mist of the cataract. In many sections, the people were terror-stricken by the awful spectacle, and supposed that the end of the world had come.

Inferior showers were seen in 1831, and 1832, and in the succeeding years, until 1839. These did not compare in brilliancy with the remarkable phenomenon of 1833. There was an interval of about 34 years between the great showers of 1799 and 1833 ; this seemed to indicate another shower in 1866 or 1867.

In November, 1866, the people of both hemispheres were literally awake to the subject. Newspapers aroused the most sluggish imagination with thrilling accounts of the scenes presented in 1799 and 1833. Extempore observatories were established at every convenient point. Watchmen were stationed, and the city bells were to be rung on the appearance of the first wandering celestial visitor. The exact night was not definitely known, but, for fear of a mistake, the 11th, 12th, and 13th were generally observed. The anxious vigils, the fruitless scannings of the sky, the disappointment, the meteors that were dimly thought to be seen,—all these were recorded in the memory of the temporary astronomers of that year.

While, however, the people of America were thus disappointed. there was enacted in England a display brilliant indeed, though inferior to the one of 1833. The staff at Greenwich Observatory counted about 8,000 meteors.

In November, 1867, the long-expected shower was seen in this country, but it failed to satisfy the public anticipation. The sky was, however, illumined with shooting stars and meteors, some of which exceeded Jupiter or Venus in brilliancy.

Number of Meteors and Shooting Stars.—Prof. Newton estimates that the average number of meteors that traverse the atmosphere daily, and which are large enough to be visible to the eye on a dark, clear night, is 7,500,000 ; and if to these the telescopic meteors be added, the number would be increased to 400,000,000. In the space traversed by the earth, there are, on the average, in each volume the size of our globe (including its atmosphere), as many as 13,000 small bodies, each one capable of furnishing a shooting star visible under favorable circumstances to the naked eye.

Annual Periodicity of the Star-Showers.—On almost any clear night, from five to seven shooting stars may be seen per hour, but in certain months they are much more abundant. Arago names the following principal dates :

April 4–11 ; 17–25.	October (about) 15.
August 9–11.	November 13–14.

Origin.—Aërolites, meteors, and falling stars are produced by small bodies—planets in miniature—revolving, like our earth, about the sun. Their orbits intersect the orbit of the earth, and if, at any time, they reach the point of crossing exactly with the earth, there is a collision. Their mass is so small, that the earth is not jarred any more than a railway train would be by a pebble thrown against it.

These small bodies may come near the earth and be drawn to its surface by the power of attraction ; or they may sweep through the higher regions of the atmosphere, and then escape its grasp ; or, finally, they may, under certain conditions, be compelled to revolve many times around the earth as satellites.

The November "meteoroids" (as these bodies are called before igniting) move at the rate of 26 miles per second in a direction nearly opposite that of the earth. They, therefore, meet our atmosphere with a relative velocity of 44 miles per second. As they sweep through the air, the friction partly arrests their motion, and converts it into heat and light. The body thus becomes visible to us. Its size and direction determine its appearance. If very small, it is consumed in the upper regions, and leaves only the luminous trail of a shooting star. If of large size, it may sweep along at a high elevation, or plunge directly toward the ground. Becoming highly heated in its course, it sheds a vivid light, while, unequally expanding, it explodes, throwing off large fragments which fall to the earth as aërolites, or continue their separate course as meteors. The cinders of the consumed portion rain down on us as fine meteoric dust.*

Meteoric Rings.—These little bodies, it is thought, do not generally revolve individually about the sun, but myriads of them are collected in a ring. When the earth passes through one of these floating girdles, a star-shower follows. This would account for their regular appearance at certain seasons of the year. The November meteoroids are not, like the August ones, uniformly distributed through the ring, but are principally collected in a swarm that has a period of 33¼ years ; hence the August shower occurs quite regularly each summer, while the great November one happens only three times in a century. The orbit of the November stream extends beyond that of Uranus. The point where it crosses the earth's orbit moves forward about 50″ per annum, and thus that star-shower occurs about a day later at each return. It takes three or four years for this

---

* Prof. Young estimates that 100 tons of meteoric matter fall upon the earth daily from outer space.

swarm to pass the node, showing that the shoal of meteoroids occupies about $\frac{1}{10}$ of its orbit. The earth in its annual revolution about the sun is supposed to encounter several hundred of these meteoric rings.

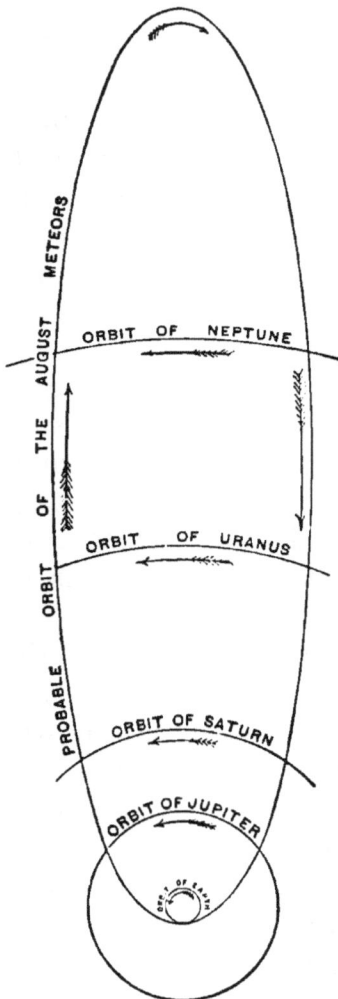

Fig. 75.

Orbit of the August Meteors.

The Physical Relation between meteoroids and comets is now generally acknowledged. The orbit of the August meteors is known to be identical with Comet III 1862 (Swift's), and that of the November 14th shower corresponds with Comet I 1866 (Tempel's). The small showers of November 24 and 27 are thought to be produced by meteors traveling in the path of the two dissevered parts of Biela's comet.

The grand problem of meteoric astronomy to-day is to identify the numerous meteoric rings, and to detect their allied comets. Being thus intimately associated, they must have a common history. Prof. Newton, the great advocate of this theory, broadly asserts that every meteoric stone was once a part of a comet, and every meteoric shower consists of broken fragments of some known or unknown comet.

Radiant Point.—The meteoroids are, of course, moving in parallel lines, but, by an optical illusion, they seem to radiate in all directions, the radiant point being in that part of the heavens which the earth is then approaching.* A star (μ) in the blade of the sickle is the point from which the stars in the November shower radiate, while one in Perseus (γ) is the radiant point of the August shower.

Height.—Herschel estimates the average height of shooting stars above the earth to be seventy-three miles at their appearance, and fifty-two at their disappearance.

Weight.—Prof. Harkness calculates that the average weight of shooting stars does not differ much from one grain.

IV. COMETS.

We come now to notice a class of bodies the most fascinating, perhaps, of any in astronomy. The suddenness with which comets flame out in the sky, the enormous dimensions of their fiery trains, the swiftness of their flight, the strange and mysterious forms they assume, their departure as unheralded as their advent,—all seem to bid defiance to law, and partake of the marvellous. Superstitious

---

* The same illusion is seen if, looking upward, we watch snow-flakes falling during a calm. Those coming directly toward our eyes seem to be motionless, and the rest to separate from them in diverging lines. This is the effect of perspective, and the "radiant point" is really the "vanishing point" of the parallel lines through which the meteors are moving. See Newcomb's Astronomy, p. 399.

fears have been excited by their appearance, and they have been looked upon in every age as

> " Threatening the world with famine, plague, and war ;
> To princes, death ; to kingdoms, many curses ;
> To all estates, inevitable losses ;
> To herdsmen, rot ; to plowmen, hapless seasons ;
> To sailors, storms ; to cities, civil treasons." *

Description.—The term comet signifies a *hairy body*. A comet consists usually of three parts ;—the *nucleus*, a bright point in the center of the *head;* the *coma* (hair), the cloud-like mass surrounding the nucleus ;

Fig. 76.

Fig. 77.

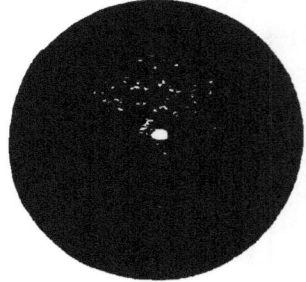

Comet without a Nucleus.   *Comet with a Nucleus.*

and the *tail*, a luminous train extending generally in a direction opposite to the sun. There are comets without the tail, and others with several tails, while some are deprived of even the nucleus. The last consist merely of a fleecy mass, known to be a comet from its orbit and rapid motion.

---

* Thus the comet of 43 B. C., which appeared just after the assassination of Julius Cæsar, was looked upon by the Romans as a celestial chariot sent to convey his soul heavenward. An old English writer observes : "Cometes signifie corruptions of the ayre. They are signes of earthquakes, of warres, of changyng kyngedomes, great dearthe of corn, yea, a common death of man and beast." Another remarks : " Experi ence is an eminent evidence that a comet, like a sword, portendeth war ; and a hairy comet, or a comet with a beard, denoteth the death of kings, as if God and nature intended by comets to ring the knells of princes, esteeming bells in churches upon earth not sacred enough for such illustrious and eminent performances."

Comets are not confined, like the planets, to the limits of the zodiac, but appear in every quarter of the heavens, and move in every conceivable direction. When first seen, the comet resembles a faint spot of light upon the dark background of the sky: as it approaches the sun the brightness increases, and the tail begins to show itself. Generally it is brightest near perihelion, and gradually fades away as it recedes, until it is finally lost, even to the telescope.*

The Time of Greatest Brilliancy depends somewhat on the position of the earth. If, as represented in the figure, the earth is at *a* when the comet, moving toward perihelion, is at *r*, the comet will appear more distinct than when it is more distant at P, although at the latter point it is really brighter. If, however, the earth is at *c* at the time of perihelion, the comet will be much more conspicuous. Again, if the earth is passing from *a* to *b* during the time

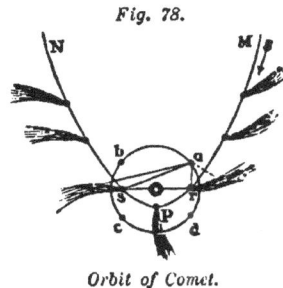

Fig. 78.

Orbit of Comet.

* While a comet remains in regions beyond the planets, where the temperature is below −140° C., its matter must be chiefly solid or liquid. On its approach to the sun, its enveloping atmosphere (if none existed, one will now be formed) will expand, and the nucleus will appear, surrounded by a blaze of light, feeble at first, but becoming more and more brilliant, and so producing the head, or coma, of the comet. Many comets do not go beyond this first phase, and, being exposed only to a moderate heat, remain telescopic. Others, piercing further the solar system, and reaching a higher temperature, develop a more abundant atmosphere. The sun, while attracting to himself the nucleus, has power to repel some of the matter of the atmosphere; how or why, we know not. Enough, that certain parts fly off as if driven by a gale, so making the tail, which increases more and more until the atmosphere is exhausted. Meanwhile, remarkable changes take place in the nucleus. Eruptions occur. Pieces are sometimes thrown off large enough to form a new comet, and showers of spark-like particles, with occasionally stony masses, fill the orbit of the comet with meteoroids.—*Schiaparelli.*

the comet is near the sun, it will appear less brilliant than if the earth were moving from c to d, as we should then be much nearer it during its greatest illumination.

Number of Comets.—Kepler remarks that "there are as many comets in the heavens as fish in the sea." Arago, basing his calculations on the number known to exist between the sun and Mercury, has estimated that there are 17,500,000 within the solar system. Of this vast number, few are visible to the naked eye, and a still less number attract observation, owing to their inferior size and brilliancy. Many are doubtless lost to our sight by being above the horizon in the daytime. During the eclipse of 1882, Lockyer, who was in Egypt to take observations, saw a brilliant comet near the sun.

Orbits of the Comets.—Comets form a part of the solar system, and are subject to the laws of gravitation. Like the planets, they revolve around the sun, though they differ in the form of their orbits. While the planets move in paths varying but little from circular, and thus never depart so far from the sun as to be invisible to us, the comets travel in extremely elongated (flattened) ellipses, so that they can be observed by us through only a small portion of their paths.

In Fig. 79 are represented the three general classes of cometary orbits. A comet traveling along an elliptical orbit, though it may pass far from the sun, will yet return within a fixed time ; one pursuing either a parabolic or hyperbolic curve cannot return, as the two sides separate from each other more and

more. Many of the comets of the first class have been
calculated, and they have repeatedly visited our portion of the heavens ; while those of the other classes,
having once visited our system, go away forever,

Fig. 79.

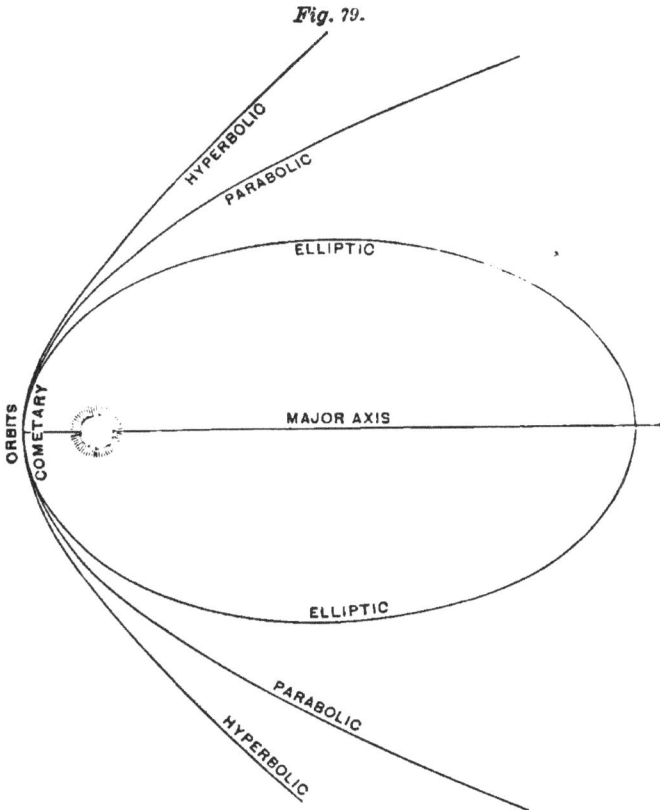

Three Forms of Cometary Orbits.

seeking perhaps in the far-off space another sun,
which in turn they will abandon as they have our own.

Calculation of a Comet's Return.—As we can c͏ᷓᵉᵗ
serve so small a proportion of the entire orbit, ᷓᵈ pre-
ed this
very difficult, indeed oftentimes impossible, to d͏ᷓ

whether it is an *hyperbola,* an *ellipse,* or a *parabola.*
A few are known to move in elliptical paths, and
their orbits have been so accurately computed that it
is possible to predict the time of their appearance.
The other comets may never return, or at least not

Fig. 80.

Projections of a few Cometary Orbits on the Plane of the Ecliptic.

for centuries hence. They may be paying our sun
their first visit ; or, if they have swept through the
solar system before, it may have been at so remote a
time that no record is preserved, even if it were not
as the creation of man. Under these circum-

stances, it is difficult to determine the place of these apparently erratic wanderers; yet, in spite of all these obstacles, some have been tracked into space far beyond the telescopic view. For example, the comet of 1844 is announced to pay a visit to the astronomers of the year of our Lord 101,844. The period of the comet of 1744, is fixed at 122,683 years.

Distance from the Sun.—Some comets at their perihelion sweep near the sun. Thus the one of 1680 came where the temperature was estimated by Newton to be about 2,000 times that of red-hot iron.* The nearest approach known is that of the comet of 1843, whose perihelion distance was but about 30,000 miles from the surface of the sun; in fact, it doubled around that body in two-hours time. (Guillemin.)† The greatest aphelion distance yet estimated is that of the comet of 1844, which is over 400,000,000,000 miles. The velocity varies, of course, with the position in the orbit. The comet of 1680 moved in perihelion at the rate of over two hundred and seventy-seven miles per second; while in aphelion its velocity is only about six miles per hour.

Density of Comets.—The quantity of matter contained in a comet is exceedingly small. Even tele scopic stars are visible through the densest part. The comet of 1770 became entangled among Jupiter's

---

* The comet of 1680 excited such terror in Europe that a medal was struck, to quiet the fears of the people. The inscription read thus: "The star threatens evil things; trust only! God will turn them to good." Newton calculated the orbit of this comet and proved that the comet moves around the sun in obedience to the law of gravity.

† The comet of 1843 excited much interest in this country since one Miller had predicted that the end of the world would come in that year; his followers imagined this comet presaged the destruction of all things.

moons, and remained there four months without in
terfering with their movements ; indeed, so far from
that, its own orbit was so much changed by their
proximity, that, from a periodical return of 5½ years,
it has not been seen since. We have good reason to
suppose that the earth, in 1861, passed through the
tail of a comet, its presence being indicated only by
a peculiar phosphorescent mist. So that even should
our earth run full-tilt against a comet, the shock
might be quite imperceptible.* Still, however
lightly we may speak of the probability of such a
collision, we must remember that there are comets
of greater solidity. Donati's, for instance, is estimated by some to be about $\frac{1}{700}$ the mass of the earth.
The concussion of such a body, moving with the
speed of a cannon-ball, would undoubtedly produce
a very sensible effect.

It is not determined whether comets shine by their
own or by reflected light. If, however, their nuclei
consist of white-hot matter, a passage through such
a furnace would be anything but desirable or satisfactory. After all the calculations of Astronomy,
our only safety lies in that Almighty Power which
traces the path and guides the course alike of
planets and comets : He, whose eye marks the fall

---

* "However dangerous might be the shock of a comet, it might be so slight that it
would only do damage to that part of the earth where it actually struck ; perhaps,
even, we might cry quits, if, while one kingdom were devastated, the rest of the earth
were to enjoy the rarities which a body coming from so far might bring to it. Perhaps
we should be very surprised to find that the *debris* of these masses that we despised
were formed of gold or diamonds ; but who would be the more astonished—we or the
comet-dwellers who would be cast upon our earth? What strange beings each would
find the other?" *Lettre sur la Comète*, par M. De Maupertuis.

Young says, "It seems, on the whole, more probable that a comet is only a cloud of
dust and vapor— a smoke-wreath—than that there is at the center any solid kernel. A
comet is a mere airy nothing."

of the sparrow, sees as well the flight of· the worlds He has created.

Variations in Form and Dimensions.—Comets appear to be subject to constant variations. They are now thought generally to decrease in brilliancy at each successive revolution about the sun. The same comet may present itself sometimes with a tail, and sometimes without. When the comet first appears, there is commonly no tail visible, and the light is faint. As it approaches the sun, however, its brightness increases, the tail shoots out from the coma, and grows daily in length and splendor. Supernumerary tails, shorter and less distinct than the principal one, dart out, but they generally soon disappear, as if from lack of material. The tail of the comet of 1843, just after the perihelion, increased in length 5,000,000 miles per day. As the tail thus extended, the nucleus was correspondingly contracted, so that this comet actually "exhausted its head in the manufacture of its own tail."

Remarkable Comets.—Among the many comets celebrated in history, we shall notice only some of those that have appeared in the present century. The *great comet* of 1811 was a magnificent spectacle.* The head was 112,000 miles in diameter; the nucleus was 400 miles; while the tail, of a beautiful fan-shape, stretched out 112,000,000 miles. "The aphelion distance of this comet is fourteen times that of Neptune, or 40,000,000,000 miles. It is announced to return in thirty centuries!" To what profound depths

---

* This was considered by the Russians to presage Napoleon's Invasion.

of space, beyond the solar system, beyond the reach
of the telescope, must such a journey extend !

Fig. 81.

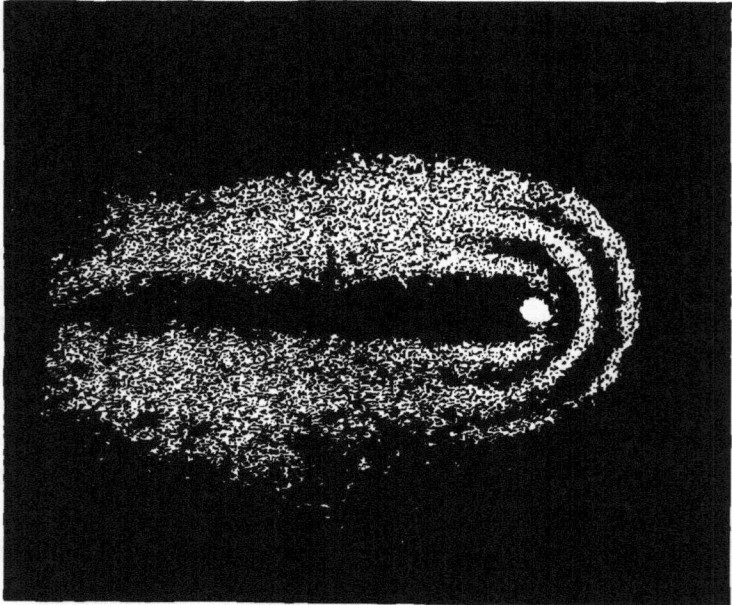

Coggia's Comet, 1874.

THE COMET OF 1835 is known as Halley's comet.
This is remarkable as being the first comet whose
period of revolution was satisfactorily established.
Dr. Halley, on examining the accounts of the great
comets of 1531, 1607, and 1682, suspected that they
were the reappearances of the same comet, whose
period he fixed at about 75 years.* He finally ven-

---

* The history of this comet, as it has been traced back by its period of seventy-five
years, is quite eventful. It was seen in England in 1066, when it was looked upon with
dread as the forerunner of the victory of William of Normandy. It was then equal to
the full moon in size. In 1456, its tail reached from the horizon to the zenith. It was

tured to predict the return of the comet at near the end of 1758 or beginning of 1759. Although Halley did not live to see his prophecy fulfilled, great interest was felt in the result. It was not destined, however, for a professional astronomer to be the

Fig. 82.

Donati's Comet.

first to detect the comet. A peasant near Dresden saw it on Christmas night, 1758.

supposed to indicate the success of Mahomet II., who had already taken Constantinople, and then threatened the whole Christian world. Pope Calixtus III., therefore, ordered extra *Ave Marias* to be repeated by everybody, and also the church bells to be rung daily at noon (whence originated the custom now so universal). A prayer was added as follows: "Lord, save us from the devil, the Turk, and the comet." In 1223, it was considered the precursor of the death of Philip Augustus of France. The first recorded appearance of Halley's comet was B. C. 130, when it was supposed to herald the birth of Mithridates.

THE COMET OF 1843 was so brilliant that it was visible in full daylight. It was so near the sun at perihelion as "almost to graze his surface."

ENCKE'S COMET has a period of only 3½ years. A most interesting discovery has been made from observations upon its motion. The comet returns each time to its perihelion about 2¼ hours earlier than the calculations indicate. Hence, Prof. Encke has been led to conjecture that space is filled with a thin, ethereal medium capable of diminishing the centrifugal force, and thus contracting the orbit of a comet.

DONATI'S COMET (1858) was the subject of universal wonder. When first discovered, in June, it was 240,000,000 miles from the earth. In August, traces of a tail were noticed, which expanded in October to about 50,000,000 miles in length. This comet, though small, has never been exceeded in the brilliancy of the nucleus and the graceful curvature of the tail. It will return in about 2,000 years.

THE "GREAT COMET OF 1882" had, soon after passing its perihelion, a nucleus as bright as a star of the 1st magnitude, and a tail 60,000,000 miles long. The aphelion of its orbit is six times further than Neptune from the sun, and the comet's period is estimated at between eight and nine centuries.

V. ZODIACAL LIGHT.

Description.—If we watch the western horizon in March or April, just after sunset, we shall sometimes see the short twilight of that season illuminated by a faint nebulous light, of a conical shape, flashing upward, often as high as the Pleiades. In September and October, at early dawn, the same appearance

Fig. 83.

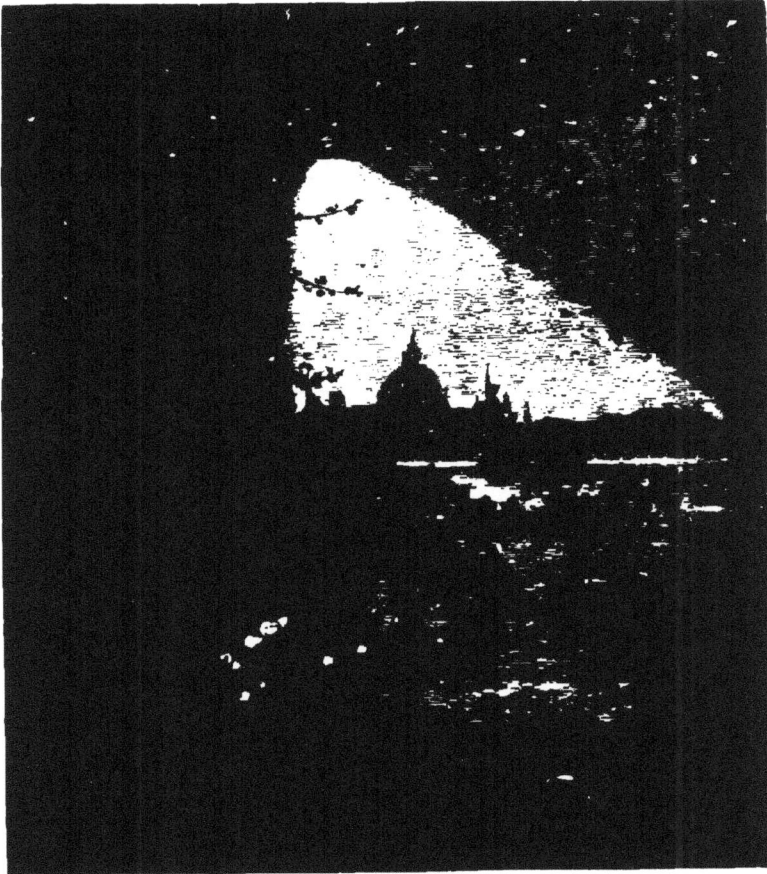

Zodiacal Light.

can be detected near the eastern horizon. The light can be seen in this latitude only on the most favorable evenings, when the sky is clear and the moon absent. Even then, it will be frequently confounded with the Milky Way or auroral lights. At the base, it is of a reddish hue, where it is so bright as often to efface the smaller stars. In tropical regions, the zodiacal light is perpetual, and shines with a brilliancy sufficient, says Humboldt, to cast a sensible glow on the opposite part of the heavens.

Origin.—The commonly-received opinion is, that it is caused by a faint, cloud-like ring, perhaps a meteoric zone, that surrounds the sun, and becomes visible to us only when the sun himself is hidden below the horizon. Others maintain that, since it has been seen in tropical regions in the east and the west simultaneously, it can be explained only on the theory of a "nebulous ring that surrounds the earth within the orbit of the moon."

PRACTICAL QUESTIONS.

1. Would the earth rise and set to a Lunarian ?
2. Could there be a transit of Neptune ?
3. Why does Mars's inner-moon rise in the west ?
4. In what part of the sky do you always look for the planets !
5. Show how it was impossible for the darkness that occurred at the time of the Crucifixion of Christ to have been caused by an eclipse of the sun.
6. Is there any danger of a collision between the earth and a comet ?
7. How are aërolites distinguished ?
8. When do we see the old moon in the west after sunrise ?
9. When do we see the moon high in the eastern sky in the afternoon before the sun sets ?
10. When is a planet morning, and when evening, star ?
11. Is the sun really hotter in summer than in winter ?
12. Why is a planet invisible at conjunction ?
13. Must an inferior planet always be in the same part of the sky as the sun ? A superior planet ?

14. Why, in summer, does the sun, at rising and at setting, shine on the north side of certain houses?

15. What effect does the volume of a planet have upon the force of gravity at its surface?

16. In what part of the heavens do we see the new moon? The old moon? The crescent moon?

17. What is the Golden Number in the almanac?

18. Why do we have more lunar than solar eclipses?

19. In what direction do the horns of the moon turn?

20. Is the "tidal-wave" an actual movement of the water?

21. Why does the sun "cross the line" in some years on March 21, and, in others, on March 22?

22. Do we ever see the sun where it really is?

23. At Edinburgh, Scotland, there are times when the sun rises at $3\frac{1}{2}$ o'clock A. M. and sets at $8\frac{1}{2}$ o'clock P. M., and the twilight lasts the entire night. When and why is this?

24. Which is the longest day of the year?

25. Is the moon nearer to us when it is at the horizon, or at the zenith?

26. How many solar eclipses would happen each year if the orbits of the sun and the moon were in the same plane?

27. Is there any heat in moonlight?

28. Can we see the moon during a total eclipse?

29. Which of the planets are repeating a portion of the earth's history?

30. How many times does the moon turn on its axis each year?

31. Can you explain the different signs used in the almanac?

32. Show how the moon is a prophecy of the earth's future.

33. Does the sun really rise and set?

34. Are the bright portions of the moon mountains or plains?

35. Which of the heavenly bodies are self-luminous?

36. Why is not a solar eclipse visible on the whole earth?

37. What is meant by the "mean distance" of a planet?

38. What keeps the earth in motion around the sun?

39. Do we ever see the sun after it sets?

40. When does the earth move the most rapidly in its orbit?

41. Have we conclusive evidence that any planet is inhabited?

42. When is the twilight the longest? The shortest? Why?

43. What is a moon?

44. To a person in the south temperate zone, where would the sun be at noon ?

45. Is it correct to say that the moon revolves about the earth, when we know that, according to the law of Physics, they must both revolve about their common center of gravity ?*

46. During a transit of Venus, do we see the body of the planet itself on the face of the sun ?

47. How many real motions has the sun ? How many apparent ones ?

48. How many real motions has the earth ?

49. Can an inferior planet have an elongation of 90° ?

50. How do we know the intensity of the sun's light on the surface of any of the planets ?

51. Why is the Tropic of Cancer placed where it is ?

52. What planets would float in water ?

53. How must the moons of Jupiter appear during their transit across the disk of that planet ?

54. "The shadow of the satellite precedes the satellite itself when Jupiter is passing from conjunction to opposition, but follows it between opposition and conjunction." Explain.

55. What facts point to the conclusion that Mars may, perhaps, have passed his planetary prime ?

56. Why may we conceive that Saturn and Jupiter are yet in their planetary youth ?

57. Show how, if the Nebular Hypothesis (p. 256) be accepted, the fashioning of a planet must require an enormous length of time.

58. Do we know the cause of gravitation ?

* "Strictly speaking, the moon does not revolve around the earth, any more than the earth around the moon ; but, by the principle of action and reaction, the center of each body moves around the common center of gravity of the two bodies. The earth being eighty times as heavy as the moon, this center is situated within the former, about three-quarters of the way from its center to its surface."—*Newcomb's Astronomy, p. 91.*

III.

THE SIDEREAL SYSTEM.

" He telleth the number of the stars; He calleth them all by their names."

PSALM cxlvii. 4.

THE SIDEREAL SYSTEM.

I. THE STARS
1. FIXED STARS NOT SEEN.
2. PARALLAX AND DISTANCE.
3. MOTION.
4. STARS ARE SUNS.
5. OUR SUN A STAR.
6. SOLAR SYSTEM IN MOTION.
7. NUMBER OF STARS.
8. SCINTILLATION.
9. MAGNITUDE.
10. CAUSE OF DIFFERENCE IN BRIGHTNESS.
11. NAMES.
12. THE CONSTELLATIONS.
13. INVENTION OF CONSTELLATIONS.
14. SIGNS AND CONSTELLATIONS NOT AGREEING.
15. PERMANENCE OF CONSTELLATIONS.
16. VALUE OF STARS.
17. ANCIENT VIEWS.
18. THREE ZONES.

II. THE CONSTELLATIONS

1. NORTHERN CIRCUMPOLAR CONSTELLATIONS, for Latitude of New York
1. HOW TRACED.
2. URSA MAJOR.
3. URSA MINOR.
    a. Description.
    b. Principal Stars.
    c. Mythological Hist.
    d. Distance of Polaris.
    e. Latitude.
4. DRACO.
5. CEPHEUS.
6. CASSIOPEIA.

2. EQUATORIAL CONSTELLATIONS.
1. HOW TRACED.
2. PERSEUS.
3. ANDROMEDA.
4. ARIES.
5. TAURUS.
6. AURIGA.
7. PISCES.
8. CETUS.
9. GEMINI.
10. ORION.
11. CANIS.
12. LEO.
13. CANCER.
14. VIRGO.
15. HYDRA.
16. CANES VENATICI.
17. BERENICE'S HAIR.
18. BOÖTES.
19. HERCULES.
20. CORONA.
21. SERPENTARIUS.
22. LIBRA.
23. SAGITTARIUS.
24. CAPRICORNUS.
25. CYGNUS.
26. LYRA.
    a. Description.
    b. Principal Stars.
    c. Mythological Hist.

3. THE SOUTHERN CONSTELLATIONS.

III. DOUBLE STARS, STAR CLUSTERS, COLORED STARS, ETC
1–6. DOUBLE STARS, COLORED STARS, VARIABLE STARS, TEMPORARY STARS, STAR CLUSTERS, NEBULÆ.
7. MAGELLANIC CLOUDS.
8. THE MILKY WAY.
9. THE NEBULAR HYPOTHESIS.

IV. CELESTIAL CHEMISTRY
1. SPECTRUM ANALYSIS.
2. SPECTROSCOPE.
3. REVELATIONS CONCERNING SUN.
4. CONCERNING STARS.
5. CONCERNING NEBULÆ.
6. CONCERNING SOLAR FLAMES.

V. TIME
1. SIDEREAL.
2. SOLAR.
3. MEAN SOLAR.
4. SUN-DIAL, ETC.

VI. CELESTIAL MEASUREMENTS
1. TO FIND DISTANCE OF PLANETS FROM SUN.
2. TO FIND MOON'S DISTANCE FROM EARTH.
3. TO FIND SUN'S DISTANCE FROM EARTH.
4. TO FIND LONGITUDE OF A PLACE, ETC.

THE SIDEREAL SYSTEM.

I. THE STARS.

IN our celestial journey we have reached Neptune, the sentinel outpost of the solar system. We are now nearly 2,800,000,000 miles from our sun. Yet we are apparently no nearer the fixed stars than when we started. They twinkle as serenely there in the far-off sky as to us here on the earth. The heavens by night, with the exception of a few changes in the planets, look familiar. Between them and us there is still a vast chasm which no imagination can bridge; a distance so immense that figures are meaningless, and we can only call it *space*,—so profound that to us it is limitless, though beyond we see other worlds twinkling, like distant lights over a waste of waters.

We never see the Stars.—This assertion seems paradoxical, yet it is strictly true. So far are the stars removed from us, that we see only the light they send, but not the surface of the worlds themselves. They are merely glittering points of light. The most powerful telescope fails to produce a sensible disk. This constitutes a marked difference between a planet and a fixed star.

The Annual Parallax of the Fixed Stars.—When speaking of this subject on page 121, we said that 186,000,000 miles, or the diameter of the earth's orbit, is the unit for measuring the parallax of the fixed stars. Yet when the stars are viewed from even these extreme points, they manifest so slight a change of place, that to estimate it is one of the most delicate feats of astronomy.

At the present time, it is considered that the star Alpha (*a*) Centauri in the southern heavens is the nearest to the earth. Its parallax is judged to be about 1″. Its distance is more than 200,000 times that of the earth from the sun, or *twenty trillions of miles.* This is probably by no means its actual distance, but merely the limit *within* which it cannot be, but *beyond* which it must be.*

These figures convey to our mind no idea of distance. Our imagination fails to grasp the thought, or to picture the vast void across which we are gazing. We remember that light moves at the rate of 186,000 miles per second. A ray at that speed would, in one day, plunge out into the abyss beyond Neptune six times the distance of that planet from the sun. Yet it must sweep on at this prodigious speed, day and night, for over 3½ years to span the gulf

* David Gill, the Royal Astronomer at the Cape of Good Hope, has recently determined the parallax of a Centauri to be 0″.75. This would make its distance 275,000 astronomical units. 275,000 × 93,000,000 miles = over 25½ trillion miles. Light would require about 4½ years to travel this enormous distance. Vega's parallax is placed at not far from 0″.2, which indicates a distance of about 1,000,000 astronomical units. Hence, Vega shines upon us from the inconceivable distance of *ninety-three trillion miles!* The parallax of Sirius has been variously estimated at from 0″.16 to 0″.38. Newcomb places this star at more than a million radii of the earth's orbit away from us, yet its light is four times as brilliant as that of any other star. The difficulty of measuring the stellar parallax may be judged from the fact that 1″ measures the angle at which a globe three-tenths of an inch in diameter would be seen when a mile away.

and reach a stopping point at the nearest fixed star.
It has been estimated that the average time required for the light of the smallest stars which are
visible to the naked eye to reach the earth is about
125 years. What, then, shall we say of those far-
distant ones, whose faint light appears as a mere
fleecy whiteness even in the most powerful telescopes? The conclusion is irresistible, that the light
we receive set out on its sidereal journey far back in
the past, perhaps before the creation of man!

Motion of the Fixed Stars.—It will aid us still
further in comprehending the immense distances of
the stars, to learn that, though they seem to be fixed,
they are moving much more swiftly than any of
the planets. Thus, Arcturus flies through space at
the astonishing rate of 200,000 miles per hour, or
nearly twice that of Mercury, and more than three
times that of the earth. Yet, through all our lifetime, we shall never be able to detect any change in
its position. "It requires three centuries for it to
move over the starry vault a space equal to the
moon's apparent diameter."

The Stars are Suns.—The vast distance at which
the stars are known to be, precludes the thought of
their shining, like the planets or the moon, by reflecting back the light of our sun. They must be self-
luminous, and are doubtless each the center of a
system of planets and satellites.

Our Sun a Star.—As we see only the suns of these
distant systems, so their inhabitants see only the sun
of our system, and that as a *small star*.

Our System in Motion.—Like all the other stars,

our sun is in motion. It is sweeping onward, with its retinue of worlds, 150,000,000 miles per year, toward a point in the constellation Hercules. The Pleiades has been thought to be the center around which this

Fig. 84.

A part of the Constellation of the Twins.

great movement is taking place, but most astronomers consider the idea as a mere speculation.

The Number of the Fixed Stars.—When we look at the heavens on a clear night, the stars seem innumerable. To count them, one would think almost as interminable a task as to number the leaves on the trees. It is, therefore, somewhat startling to learn that the entire number visible to the most piercing eyesight does not exceed 6,000, while few can discern more than 4,000.* The number, however, which may be seen with a telescope is marvellous. In Fig. 84, is shown a portion of the heavens where the naked eye sees but six stars. Could we examine the same region of the sky with more powerful instruments, new constellations would doubtless be descried in the infinite depths of space.

Scintillation.—The twinkling of the fixed stars is due to what is termed in Physics the "Interference of Light." The air, being unequally dense, warm, and moist in its various strata, transmits very irregularly the different colors of which white light is composed. Now one color prevails over the rest, and now another, so that the star appears to alter its hue incessantly. As the purity and density of the air vary, the twinkling of the stars also changes, and, therefore, it is always greatest near the horizon.†

Magnitude of the Stars.—As the telescope reveals no disk of even the nearest stars, we know nothing of their comparative size. The finest spider's thread, placed at the focus of the instrument, hides the star from the eye. When the moon passes in front of a

---

* This illusion may be easily explained, when we remember how the impression of a bright light remains upon the retina, as in the whirling of a firebrand.

† Humboldt says that at Cumana, in South America, where the air is remarkably pure and uniform in density, the stars cease to twinkle after they have risen 15° above the horizon. This gives to the celestial vault a peculiarly calm and soft appearance.—It should be noticed that interference occurs only when the light emanates from a point. A body that subtends a visual angle, i. e., has a sensible disk, like a planet, cannot twinkle.

star, the occultation is instantaneous, and not gradual, as in the case of the planets. Classification depends, therefore, merely upon their relative brightness. The most conspicuous are termed *stars of the first*

Fig. 85.

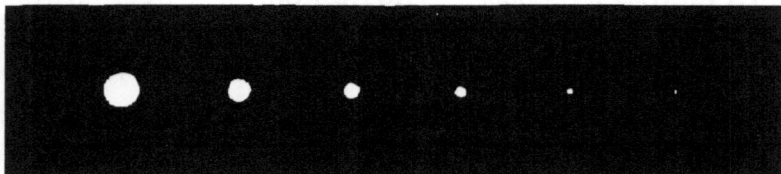

magnitude; of these there are about twenty. The number of second-magnitude stars in the entire heavens is sixty-five; of the third, about 200; of the fifth, 1,100; of the sixth, 3,200; of the seventh, 13,000; of the eighth, 40,000; and of the ninth, 142,000. Few persons can see smaller stars than those of the fifth or sixth magnitude.

The Difference in the Brightness of the stars may result from a difference in their distance, size, or intrinsic brightness. Hence it follows that the faintest stars may not be the most distant from the earth.

Names of Stars.—Many of the brightest stars received proper names at an early date; as Sirius, Arcturus. The chief stars of each constellation are distinguished by the letters of the Greek alphabet;

THE GREEK ALPHABET.

A	α	Alpha	I	ι	Iota	P	ρ	Rho
B	β	Beta	K	κ	Kappa	Σ	σ	Sigma
Γ	γ	Gamma	Λ	λ	Lambda	T	τ	Tau
Δ	δ	Delta	M	μ	Mu	Υ	υ	Upsilon
E	ε	Epsilon	N	ν	Nu	Φ	φ	Phi
Z	ζ	Zeta	Ξ	ξ	Xi	X	χ	Chi
H	η	Eta	O	o	Omicron	Ψ	ψ	Psi
Θ	θ	Theta	Π	π	Pi	Ω	ω	Omega

the brightest one being usually called Alpha (*a*), the next Beta (*β*), etc.,—the name of the constellation, in the genitive case, being put after each. Ex., *a* Arietis, *β* Lyræ.*

Star catalogues are issued, containing the stars arranged in the order of their Right Ascension, and numbered for convenience of reference. Argelander's Charts have 324,188 stars marked in the northern hemisphere.

The Constellations.—From the earliest ages, the stars have been arranged in constellations, for the purpose of more readily distinguishing them. Some of these groups were named from their supposed resemblance to certain figures, such as perching birds, pugnacious bulls, or contorted snakes, while others do honor to the memory of classic heroes.

> "Thus monstrous forms, o'er heaven's nocturnal arch,
>   Seen by the sage, in pomp celestial march ;
>   See Aries there his glittering bow unfold,
>   And raging Taurus toss his horns of gold ;
>   With bended bow the sullen Archer lowers,
>   And there Aquarius comes with all his showers ;
>   Lions and Centaurs, Gorgons, Hydras rise,
>   And gods and heroes blaze along the skies."

With a few exceptions, the likeness is purely fanciful. Not only are the figures uncouth, and the origin often frivolous, but the boundaries are not distinct. Stars occur under different names ; while one constellation encroaches upon another.† Though,

---

* This means *a* of Aries, *β* of Lyra ; the genitive case in Latin being equivalent to the preposition *of*.

† Chambers well remarks, "Aries should not have a horn in Pisces and a leg in Cetus, nor should 13 Argôs pass through the Unicorn's flank into the Little Dog. 51 Camelopardali might with propriety be extracted from the eye of Auriga, and the rils of Aquarius released from 46 Capricorni."

however, the constellations are thus rude and imperfect, there seems little hope of any change. Age
gives them a dignity that insures their perpetua-
tion.

The Invention of the Constellations goes back into
ages of which no record remains. By some it has
been ascribed to the Greeks. When the signs of the
zodiac were named, they doubtless coincided with
the constellations. Aries (the ram) was so called
because it rose with the sun in the spring-time, and
the Chaldean shepherds named it from the flocks,
their most valued possession. Then followed, in
order, Taurus (the bull) and Gemini (the twins),
called from the herds, which were esteemed next in
value. At the summer solstice, the sun appears to
stop, and, crab-like, to crawl backward; hence the
name Cancer (the crab). When the sun is in Leo,
the brooks being dry, the lion leaves his lurking-
place and becomes a terror to all. Virgo comes
next, when the virgins glean in the summer harvest.
At the autumnal equinox, the days and nights are
equally balanced, and this is beautifully represented
by Libra (the scales). The vegetation decays in the
fall, causing sickness and death; the Scorpion, which
stings as it recedes, is suggestive of this Parthian
warfare. Sagittarius (the archer) tells of the hunting month. Capricornus (the goat, which delights
in climbing lofty precipices) denotes how at the
winter solstice the sun begins to climb the sky on
his return north. Aquarius (the water-bearer) is a
natural emblem of the rainy season. Pisces (the
fishes) is the month for fishing.

Signs and Constellations do not Agree. — By the precession of the equinoxes, as we have before described on page 106, the signs have fallen back along

Fig. 86.

The Signs and Constellations, as they now Compare in the Heavens, the former having fallen back, and the latter apparently advanced, 30° each.

the ecliptic about 30°, so that those stars which were, during the infancy of astronomy, in the sign Aries (♈) are now in Taurus (♉), and those which were in the sign Pisces (♓) are now in Aries (♈).*

Permanence of the Constellations.—The general appearance of the constellations and the figures

* If the teacher will put a pin at the center of Fig. 86, and then draw a sharp knife between the signs and the constellations, so as to detach the middle of the cut, and cause the inner part to revolve, the signs may be turned before any constellation, and thus this change be clearly apprehended.

which the stars form are due to the position we occupy. Could we cross the gulf of space beyond Neptune, the stars now so familiar to us would look strangely enough in their new groupings. As one in riding through a forest sees the trees apparently increase in size and open up to view before him, while they decrease in size and close in behind him, forming clusters and groups which constantly change as he passes along, so, as our earth travels with the solar system on its immense sidereal journey, the stars will gradually grow larger and brighter in front, while those behind us will appear smaller and dimmer.

Since, in addition to this, the stars themselves are in motion with varying velocity and in different directions, the constellations must change still more rapidly, so as ultimately to transform the appearance of the heavens. In time, the "Bands of Orion" will be loosened, and the "Seven Sisters" will glide apart. Such are the distances, however, that, although these movements have been going on constantly, no variation has occurred, since the creation of man, that is perceptible, save to the watchful astronomer. Nothing in nature is so invariable as the stars. They are the standards of time. Myriads of years must elapse before new maps of the constellations will be required.

Value of the Stars in Practical Life.—"The stars are the landmarks of the universe." They seem to be placed in the heavens by the Creator, not alone to elevate our thoughts and expand our conceptions of the infinite and eternal, but to afford us, amid the constant fluctuations of our own earth, something unchangeable and abiding. Every object about us is constantly shifting, but over all shine the "eternal stars," each with its place so accurately marked, that to the astronomer and the geographer no deception is possible. To the mariner, the heavens become a dial-plate, the figures on its face set with glittering stars, along which the moon travels as a shining hand that marks off the hours with an accuracy no watch can ever rival. Standing on the deck of his vessel, far out at sea, a single observation of the sun or the stars decides his location in the waste of waters as accurately as if he were at home, and had caught sight of some old landmark he had known from his boyhood. In all the intricacies of surveying, the stars furnish the only immutable guide. Our clocks vainly strive to keep time with the celestial host. Thus, by an evident plan of the Creator, even in the most common affairs of life, are we compelled to look for guidance from the shifting objects of earth up to the heavens above.

ANCIENT VIEWS.—Anaximenes (500 B. C.) thought that the stars were for ornaments, and were nailed like bright studs into the crystalline sphere. Anaxagoras considered that they were stones whirled up from the earth by the rapid motion of the ether, and that its inflammable properties set them on fire and caused them to shine as stars. Some schools of the Grecian philosophers—the Stoics, Epicureans, etc.— believed that they were celestial fires kept alive by matter that constantly streamed up to them from the center of the heavens. The stars were at one time said to feed on air ; at another, to be the breath-ing holes of the universe.

Three Zones of Stars.—If we recall what was said on page 90, concerning the paths of the stars and the appearance of the heavens at different seasons of the year, we shall see that the constellations are nat-urally divided into three zones. The *first* embraces those which are visible through the entire year ; the *second,* those whose paths can be seen only in part on any given night ; and the *third,* those whose paths just graze our southern horizon, or never pass above it.

II. THE CONSTELLATIONS.

I. The Northern Circumpolar Constellations are visible in our latitude every night. They may be easily traced by holding the book up toward the northern sky in such a way that Polaris and the Big Dipper on the map and in the heavens agree in posi-tion, and then locating the other constellations by comparison.

As the stars revolve about Polaris, their places will vary with every successive night through the year. The cut represents them as they are seen at midnight of the winter solstice. At 6 **P. M.** of that day, the right-hand side of the map should be held downward, and the Big Dipper will be directly below the north star. At 6 **A. M.**, the left-hand side should be at the bottom, and the Dipper will be above Polaris. · From day to day, this aspect will change, each star coming a little earlier to the meridian, or to its position on the preceding night. The rate of this progression is six hours, or 90°, in three months.

(*Map No. 1.*) *Fig. 87.*

Northern Circumpolar Constellations.

Ursa Major is represented under the figure of a great bear. It contains 133 stars visible to the naked eye. This constellation has been celebrated among all nations. It is remarkable that the shepherds of Chaldea in Asia and the Iroquois Indians of America gave to it the same name.

PRINCIPAL STARS.—A noticeable cluster of seven stars—six of the second and one of the fourth magnitude—forms what is familiarly termed the *Dipper*. In England it is styled Charles's Wain, from a fancied resemblance to a wagon drawn by three horses tandem. Mizar (ζ) has a minute companion, Alcor, which Humboldt tells us could be rarely seen in Europe. A person with good eyesight may now readily detect it. Megrez (δ), at the junction of the handle and the bowl, is to be marked particularly, since it lies almost exactly in the colure passing through the autumnal equinox. Dubhe and Merak are termed the Pointers, because they point out the polar star. The bear's right fore-paw and hind-paw* are each marked by two small stars, as shown in the cut; a similar pair nearly in line with these denote the left hind-paw (see ξ, Fig. 90).

MYTHOLOGICAL HISTORY.—Diana had a beautiful attendant named Callisto. Juno, the queen of heaven, becoming jealous of the maid, transformed her into a bear.

> " The prostrate wretch lifts up her head in prayer,
> Her arms grow shaggy, and deformed with hair ;
> Her nails are sharpened into pointed claws,
> Her hands bear half her weight and turn to paws.
> Her lips, that once would tempt a god, begin
> To grow distorted in an ugly grin.
> And lest the supplicating brute might reach
> The ears of Jove, she was deprived of speech.
> How did she fear to lodge in woods alone,
> And haunt the fields and meadows once her own !
> How often would the deep-mouthed dogs pursue,
> Whilst from her hounds the frighted hunters flew."

* It is well to notice that Dubhe and Merak are about 5° apart ; Dubhe and Benetnasch are about 25° apart ; the paws of the Bear are 15° apart ; while Polaris is about 30° distant.

Some time afterward, Callisto's son, Arcas, being out hunting, pursued his mother, and was about to transfix her with his uplifted spear, when Jupiter in pity transferred them both to the heavens, and placed them among the constellations as Ursa Major and Ursa Minor.

Ursa Minor is represented under the figure of a small bear. It contains twenty-seven stars, of which only three are of the third, and four of the fourth magnitude.

PRINCIPAL STARS.—A cluster of seven stars forms the *Little Dipper*. Three of them are small, and are seen with difficulty. Polaris, at the extremity of the handle, has been known from time immemorial as the North Polar Star. Until the mariner's compass came into use, it was the star

> " Whose faithful beams conduct the wandering ship
> Through the wide desert of the pathless deep."

Polaris does not mark the exact position of the pole, since that is about $1\frac{1}{2}°$ toward the Pointers. This distance will gradually diminish,* until in time (2120 A. D.) it will be only $\frac{1}{2}°$: then it will increase again, until, in the lapse of ages, 12,000 years hence, the brilliant star Vega (α Lyræ) will fulfill the office of polar star for those who shall then live on the earth.†

THE DISTANCE OF POLARIS is so great, that, though the star is moving through space at the rate of ninety

---

* Five stars of the Dipper itself are drifting away from the sun, at the rate of 17 miles per second, seeming to form a family or group by themselves. Proctor's Easy Star Lessons gives charts representing the appearance of the Dipper for 100,000 years.

† Of the nine Pyramids which are standing at Gizeh, Egypt, six have openings facing the north. These lead to straight passages which descend at a uniform angle of about 26° and are parallel with the meridian. If we suppose a person, 4,000 years ago, standing at the lower end of one of these passages, and looking out, his eye would strike the sky near the star Thuban, which was then the polar star. The supposed date of the building of these Pyramids (the Great Pyramid, 2123 B. C.) agrees with that epoch, and naturally suggests that the builders had some special design in this peculiar construction.

miles per minute, this tremendous speed is imperceptible to us. It requires nearly fifty years for its light
to reach the earth; so that, when we look at Polaris,
we know that the ray which strikes our eye set out
on its journey through space half a century ago.
We cannot state positively that the star is now in
existence, since if it were destroyed to-day it would
be fifty years before we should miss it.*

CALCULATION OF LATITUDE FROM POLARIS.—By an
observer at the equator, Polaris is seen at the horizon.
If he goes north, the horizon is depressed, and Polaris
seems to rise in the heavens. When it has reached
the height of a degree, the observer is said to have
passed over a degree of latitude on the earth's surface. As he moves further north, the polar star continues to ascend; its distance above the horizon
denoting the latitude of each place in succession,
until at the north pole, if one could reach that point,
Polaris would be seen directly overhead.

Draco is represented under the figure of a long
sinuous serpent, stretching between Ursa Major and
Ursa Minor, nearly encircling the latter constellation,
and finally reaching out its head almost to the body
of Hercules.

PRINCIPAL STARS.—Four small stars form a quad-
rilateral figure at the head; a fifth, of the fourth
magnitude, which is scarcely visible, marks the end
of the nose; several scattered groups and little
triangles of small stars denote the position of the
various coils of the body; thence, an irregular line
of stars traces the dragon's tail around between Ursa

* Some recent observations seem to reduce this to 42 years.

Major and Ursa Minor.   Thuban, lying midway between γ of the Little Dipper and ζ of the Big Dipper, is noted as the polar star of forty centuries ago.

MYTHOLOGICAL HISTORY.—Jupiter had carried off Európa.   Agénor, her father, sent her brother Cadmus in pursuit of his lost sister, bidding him not to return until he was successful in his search.   After a time, Cadmus, weary of his wanderings, inquired of the oracle of Apollo concerning the fate of Europa.   He was told to cease looking for his sister, to follow a cow as a guide, and when she rested, there to build a city.   Hardly had Cadmus stepped out of the temple, when he saw a cow slowly walking along.   He followed her until she came upon the broad plains where Thebes afterward stood.   Here she stopped.   Cadmus, wishing to offer a sacrifice to Jupiter in gratitude for the delightful location, sent his servants to seek for water.   In a dense grove near by, was a fountain guarded by a fierce dragon (*Draco*), and sacred to Mars.   The Tyrians, approaching this and attempting to dip up some water, were attacked, and many of them killed, by the enormous serpent, whose head overtopped the tallest trees. Cadmus, becoming impatient, went in search of his men, and, on arriving at the spring, saw the sad disaster.   He forthwith fell upon the monster, and after a severe battle succeeded in slaying him.   While standing over his conquered foe, he heard a voice from the ground bidding him take the dragon's teeth and sow them.   He obeyed.   Scarcely had he finished when the earth began to move and the points of spears to prick through the surface.   Next, nodding plumes shook off the clods, and the heads of armed men protruded.   Soon a great harvest of warriors covered the entire plain.   Cadmus, in terror at the appearance of those giants, whom he termed Sparti (*the sown*), prepared to attack them, when suddenly they turned upon themselves, and never ceased their warfare till only five of the crowd survived.   These, making peace with one another, joined Cadmus, and assisted him in building the City of Thebes.

Cepheus is represented as a king in regal state, with a crown of stars on his head, while he holds in his hand a scepter which is extended toward his wife, Cassiopeia.   The constellation contains thirty-five stars visible to the naked eye.

PRINCIPAL STARS.—The brightest star is Alderamin (a), in the right shoulder. Alphirk (β), in the girdle, is at the common vertex of several triangles, which point out respectively the left shoulder (ι), the left knee (γ), and the right foot. The head, which lies in the Milky Way, is marked by a little triangle of three stars. This forms, with a, β, and ι, quite a regular quad-rilateral figure. A bright star of the fifth magnitude. close to Polaris, points out the left foot.

Cassiopeia* is represented as a queen seated on her throne. On her right, is the king; on her left, Perseus, her son-in-law; above her, Andromeda, her daughter. The constellation contains sixty-seven stars visible to the naked eye.

PRINCIPAL STARS.—A line drawn from Megrez (δ), in Ursa Major, through Polaris and continued an equal distance, will strike Caph (β) in Cassiopeia. This star is noticeable as marking, with the others named, the equinoctial colure, and as being on the same side of the true pole as Polaris. The principal stars form the figure of an inverted chair, which is very striking and may be easily traced.

II. Equatorial Constellations.—The constellations we shall now describe lie south of the circumpolar groups. Only a portion of their paths is above our horizon. In using the maps, the observer is supposed to stand with his back toward Polaris, and to be look-ing directly south. Commencing with the constella-tion Perseus, so intimately connected with the other

---

* For the mythological history of Cassiopeia, see Perseus and Andromeda. The names of the principal stars in the Chair make a mnemonic word.—*βαγ δε,' bugle.* The student can often form such an association of the letters, and will find the device an aid to his memory. (Compare Virgo, page 230.)

members of the royal family just described, we pass eastward in our survey, and notice the various constellations as they slowly defile in silent march across the sky.

The first map represents the constellations on or near the meridian at nine o'clock in the evening of the winter solstice. On the evening of the autumnal equinox, the left-hand side of the map should be turned downward toward the eastern horizon. On the evening of the vernal equinox, the right-hand side should be turned to the western horizon. At these different times, the stars, though keeping their relative positions, will be diversely inclined to the horizon. As the stars apparently move westward at the rate of 15° per hour, the time of the evening determines what stars will be visible, and also their distances above the horizon.

Perseus is represented as brandishing an enormous sword in his right hand, while in his left he holds the head of Medusa. The constellation comprises eighty-one stars visible to the naked eye.

PRINCIPAL STARS.—The most prominent figure is called the *Segment of Perseus.* It consists of several stars arranged in a line curving toward Ursa Major. Algenib (*a*), the brightest of these, is of the second magnitude. Algol (p. 242), in the midst of a group of small stars, marks the head of Medusa. Between the bright stars of Perseus and Cassiopeia, is a beautiful star-cluster visible to the naked eye.

MYTHOLOGICAL HISTORY.—Perseus, from whom this constellation was named, was the son of Jupiter and Danaë. His grandfather, Acrisius, having been informed by the oracle that his grandson would be the instru-

(Map No. 2)—Fig. 88.

ment of his death, put the mother and child in a coffer and set them adrift on the sea. Fortunately, they floated near the island Seriphus, where they were rescued and kindly treated by a brother of Polydectes, king of the country. When Perseus had grown up, he was ordered by Polydectes to bring him, as a marriage gift, the head of Medusa. Now Medusa was once a beautiful maiden, who dared to compare her ringlets with those of Minerva; whereupon, the goddess changed her locks into hissing serpents, and made her features so hideous, that she turned to stone every living object upon which she fixed her Gorgon gaze. Perseus was at first overpowered at the thought of undertaking this enterprise; but Mercury promised to be his guide, and to furnish him with his winged shoes; Minerva loaned him her wonderful shield, that was bright as a mirror; and the Nymphs gave him, in addition, Pluto's helmet, which made the bearer invisible. Thus equipped, Perseus mounted into the air and flew to the ocean, where he found the three Gorgons, of whom Medusa was one, asleep. Fearing to gaze in her face, he looked upon the image reflected in Minerva's shield, and with his sword severed Medusa's head from her body. The blood gushed forth, and with it the winged steed *Pegasus*. Grasping the head, Perseus flew away. The other Gorgons awaking, pursued him, but he escaped their search by means of Pluto's helmet. As he flew over the wilds of Libya, in his aerial route, the blood dripping from the gory head of the monster produced the innumerable serpents for which that country was afterward noted.

Andromeda is represented as a beautiful maiden chained to a rock.

PRINCIPAL STARS.—Algenib and Algol in Perseus form, with Almach (γ) in the left foot of Andromeda, a right-angled triangle opening toward Cassiopeia. This figure is so perfect, that the stars may be easily recognized. The girdle is pointed out by Merach (β), and two other stars which form a line slightly curving toward the right foot. The breast is denoted by a very small triangle composed of three stars,—δ of the fourth magnitude, another of the fifth magnitude just south, and an exceedingly minute star a little at the west. Alpheratz (α), in the head of Andromeda, belongs also to *Pegasus*. This star, with three others —Algenib (γ), Markab (α), and Scheat (β),—all of the second magnitude, constitute the *Great Square of Pegasus*. The brightest stars of these two constellations form a figure strikingly like the Big Dipper. Algenib and Alpheratz lie in the equinoctial colure which passes through Caph.

MYTHOLOGICAL HISTORY. —Cassiopeia had boasted that her daughter Andromeda was fairer than the Sea-nymphs. They appealed, in great indignation, to Neptune, who sent a sea-monster (*Cetus*) to devastate the coast of Ethiopia. To appease the deities, her father Cepheus was directed by the oracle to bind his daughter to a rock, to be devoured by Cetus. Perseus, returning from the destruction of Medusa, saw Andromeda in her forlorn condition. Struck by her beauty and tears, he offered to liberate her at the price of her hand. Her parents joyfully consented, and, in addition, offered a royal dower. Perseus slew the terrible monster, and freeing Andromeda, restored her to her parents. All the prominent actors in this scene were honored with seats among the constellations. The Sea-nymphs, it is said, in petty spite of Cassiopeia, prevailed that she should be placed where for half of the time she hangs with her head downward,—a fit lesson of humility. *Cepheus*, her husband, shares in her punishment.

Aries, the *ram*, was anciently the first constellation of the zodiac. It is now the *first sign*, but the *second constellation*. On account of the precession of the equinoxes, the constellation Pisces occupies the first sign.

PRINCIPAL STARS.—The most noted star is *a* Arietis (Alpha of Aries, more commonly called simply Arietis), in the right horn. This lies near the path of the moon and is one of the stars from which longitude is reckoned. A line drawn from Almach to Arietis will pass through a beautiful figure of three stars called *The Triangles.*

MYTHOLOGICAL HISTORY.—Phrixus and Helle were the children of Athamas, king of Thessaly. Being persecuted by Ino, their step-mother, they were compelled to flee for safety. Mercury provided them a ram which bore a golden fleece. The children were no sooner placed on his back than he vaulted into the heavens. In their aerial journey, Helle becoming dizzy fell off into the sea, which was afterward called the Hellespont, now the Dardanelles. Phrixus having reached Colchis in safety, offered the ram in sacrifice to Jupiter, and gave the golden fleece to Aetes, his protector. The Argonautic expedition in pursuit of this golden fleece, by Jason and his followers, is one of the most romantic of mythological stories. It is, undoubtedly, a fanciful account of the first important maritime expedition. Rich spoils were the prizes to be secured.

Taurus consists of the head and shoulders of a *bull*, which is represented in the act of plunging at Orion.

PRINCIPAL STARS.—The *Hyades*, a beautiful cluster

in the head, forms a distinct V. The brightest of these is Aldebaran, a fiery red star of the first magnitude.* The *Pleiades* (Job, xxxviii, 31), or the Seven Sisters, is the most conspicuous group in the sky (p. 206). It contains a large number of stars, six of which are visible to the naked eye. There were said to have been seven anciently, but that Electra left her place in order not to behold the ruin of Troy, which was founded by her son Dardanus. Other myths relate that the *"Lost Pleiad"* was Merope, who married a mortal. Alcyone is the brightest Pleiad. El Nath (*β*) and *ζ* point out the horns of Taurus.

MYTHOLOGICAL HISTORY.—This is the animal whose form Jupiter assumed when he bore off Europa. The Pleiades were the daughters of Atlas, and Nymphs of Diana's train. They were distinguished for their unblemished virtue and mutual affection. The hunter *Orion* having pursued them one day, in their distress they prayed to the gods, when Jupiter, in pity, transferred them to the heavens.

Auriga, the *Charioteer* or *Wagoner*, is represented as a man resting one foot on a horn of Taurus, and holding a goat and kids in his left hand and a bridle in his right.

THE PRINCIPAL STARS are arranged in an irregular five-sided figure. Capella, the goat-star, is of the first magnitude. It travels in its orbit 1,800 miles per minute; seventy years—a long lifetime—are required for its light to reach the earth. Near by is a tiny triangle, formed of three small stars, called the *Kids*. Menkalinan (*β*) is in the right shoulder, *θ* in the right

* Aldebaran is estimated to move through the heavens at the rate of 55 miles per second. (See pp. 205, 261.) A number of the bright stars between Aldebaran and the Pleiades have a common motion of about 10″ per century toward the east.

hand, ɜ (common to Auriga and Taurus) the right foot, and ɩ the left foot. Capella. β, and δ (a star in the head) form a triangle. The origin of this constellation is unknown.

Pisces, the *fishes,* is represented by two fishes tied together by a long ribbon. It consists of small stars, which can be traced only upon a clear night, and in the absence of the moon.

Cetus. the *whale,* is a huge sea-monster, slowly ploughing his way eastward, midway between the horizon and the zenith. It may easily be found, on a clear night. by means of the numerous figures given in the map.

(Map No. 3)—Fig. 89.

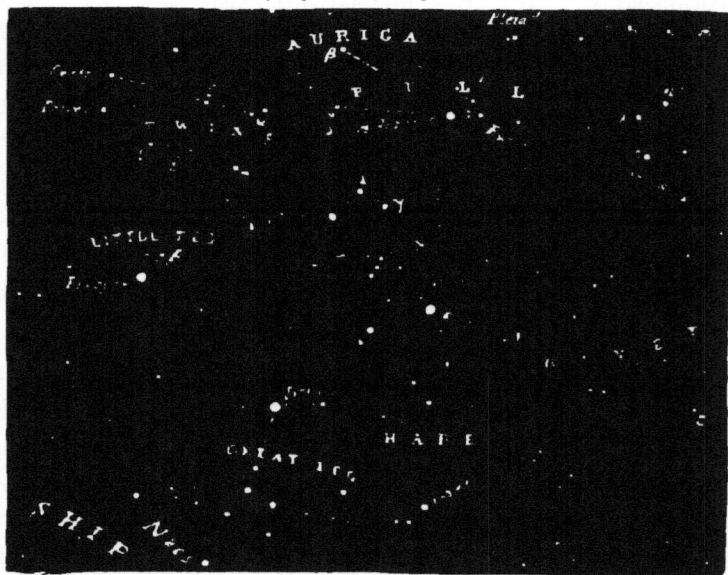

Gemini, the *Twins.* represents the twin brothers Castor and Pollux.

* "Castor is resolved by the telescope into two stars, whose angular distance from each other is 5″— the angle that one inch would subtend 1,146 yards off."—BALL.

THE PRINCIPAL STARS are Castor* and Pollux, which
are of the first and second magnitudes. The latter
is one of the stars from which longitude is reckoned
by means of the Nautical Almanac. The constellation is clearly distinguished by two nearly parallel
rows of stars, that by a slight effort of the imagination may be extended into the constellations Taurus
and Orion.

MYTHOLOGICAL HISTORY.—Castor and Pollux were noted,—the former
for his skill in training horses, the latter for boxing. They were tenderly
attached to each other, and were inseparable in their adventures. They accompanied Jason on the Argonautic expedition. A storm having arisen
during this voyage, Orpheus played on his wonderful lyre and prayed to the
gods; whereupon the tempest was stilled, and star-like flames shone upon
the heads of the twin-brothers. Sailors, therefore, considered them as
patron deities,* and the balls of electric flame seen on masts and shrouds,
now called St. Elmo's fire, were named after them. Afterward, Castor was
slain. Pollux being inconsolable, Jupiter offered either to take him up to
Olympus, or to let him share his immortality with his brother. Pollux pre-
ferred the latter, and so the brothers pass alternately one day under the
earth, and the next in the Elysian Fields. Not only did sailors thus con-
fide in their watch over navigation, but soldiers believed them to return,
mounted on snow-white steeds and clad in rare armor, to take part in the
hard-fought battle-fields of the Romans.

> "Back comes the chief in triumph,
>   Who in the hour of fight
> Hath seen the great Twin Brethren,
>   In harness on his right.
> Safe comes the ship to haven,
>   Through billows and through gales,
> If once the great Twin Brethren
>   Sit shining on the sails."— *Lays of Ancient Rome.*

Orion is represented under the figure of a hunter
assaulting Taurus. He has a sword in his belt, a

* We remember that Paul sailed for Italy in a ship whose sign was Castor and Pollux.
—Acts, xxviii. 11.

club in his right hand, and the skin of a lion in his left. This is one of the most clearly defined and conspicuous constellations in the heavens.

PRINCIPAL STARS.—Four brilliant stars, in the form of a parallelogram, mark the outlines of Orion. Betelgeuse, a beautiful ruddy star of the first magnitude, is in the right shoulder; Bellatrix (γ), of the second magnitude, is in the left shoulder; Rigel, of the first magnitude, is in the left foot; and Saiph ($\varkappa$), of the third magnitude, is in the right knee. Two small stars near λ form with it a small triangle, which is itself the vertex of a larger triangle composed of λ, γ, and Betelgeuse. Near the center of the parallelogram are three stars forming the *Belt of Orion*. This group is also called the Bands of Orion (Job, xxxviii, 31), Jacob's rod, and the Yard. It received the last name because it forms a line 3° long, divided in equal parts by a star in the center. These divisions are useful for measuring the distances of the stars. Running from the belt southward, is an irregular line of stars which marks the sword; west of Bellatrix is a curved line denoting the lion's skin. South of Orion are four stars forming a beautiful figure styled *The Hare*.

MYTHOLOGICAL HISTORY.—Orion was a famous hunter. Becoming enamored of Merope, he desired to marry her. Œnopion, her father, opposing the choice, put out the eyes of the unwelcome suitor. The blinded hero followed the sound of a Cyclop's hammer until he came to Vulcan's forge. Vulcan, taking pity, instructed Kedalion to conduct him to the abode of the sun. Placing his guide on his shoulder, Orion proceeded to the east, and at a favorable place

> "Climbing up a narrow gorge,
> Fixed his blank eyes upon the sun."

The healing beams restored him to sight. As a punishment for having profanely boasted that he was able to conquer any animal the earth could produce, he was bitten in the heel by a scorpion. Afterward, Diana placed him among the stars; where *Sirius* and *Procyon*, his dogs, follow him, the *Pleiades* fly before him, and far remote is the *Scorpion*, by whose bite he perished.

Canis Major and *Canis Minor* contain each a single star of the first magnitude, Sirius, and Procyon.* These two, with Betelgeuse, Phaet in the Dove, and Naos in the Ship, form a huge figure known as the Egyptian X. Sirius, the dog-star, is the most brilliant star in the heavens. It is receding from the earth at the rate of 20 miles per second (Huggins). Seventeen years are required for its light to reach us.† (See note, p. 308.)

Leo is represented as a rampant lion. It is one of the most beautiful constellations in the zodiac.

THE PRINCIPAL STARS are arranged in the form of a sickle. Regulus, in the handle, is a brilliant star of the first magnitude. It is one of the stars from which longitude is reckoned. It is almost exactly in the ecliptic. Zosma (δ) lies in the back of the lion, θ in the thigh, and Denebola, a star of the second magnitude, in the brush of the tail.

Cancer includes the stars that lie irregularly scattered between Gemini, Head of Hydra, Procyon, and Leo. In the midst of these, is a luminous spot, called Præsepe, or the Bee-hive, which an ordinary glass will resolve into stars.

* Procyon, like Sirius, was formerly considered a star of evil omen, and as bringing bad weather. "Who that is learned in matters astronomical," said Digges, the astrologer, "noteth not the great effects at the rising of the star called the Litel Dogge."

† In 1862, Alvan G. Clark, son of the famous telescope-maker, discovered a companion of Sirius, "distant from the star 28 times the Sun's distance from the Earth."

Virgo is represented as a beautiful maiden with folded wings, bearing in her left hand an ear of corn.

THE PRINCIPAL STAR, Spica, in the ear of corn, is of the first magnitude, and is used for determining longitude at sea. Denebola, Cor Caroli (*a*), Arcturus, and Spica form a figure about 50° in length, called

(Map No. 4)- Fig. 30.

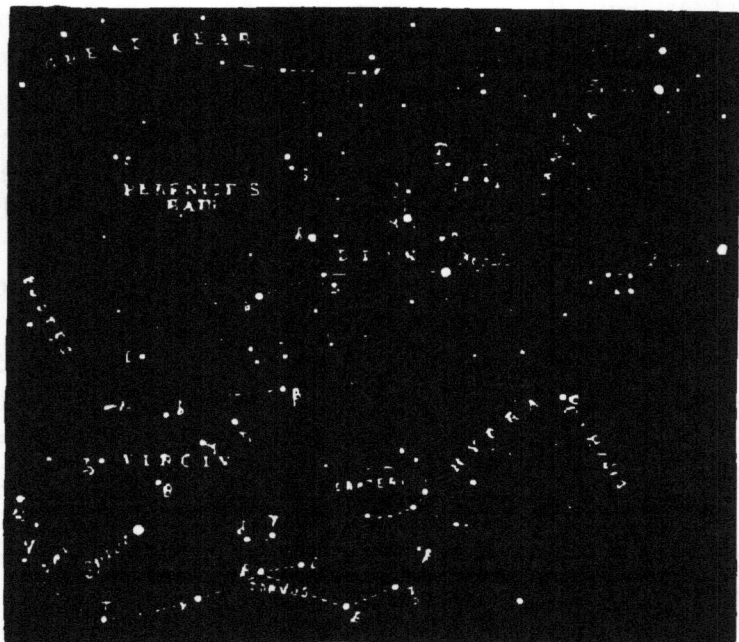

the Diamond of Virgo. Five third-magnitude stars, ε, δ, γ, η, β, (the mnemonic word is bēgde) make a corner known among the Arabian astronomers as "The retreat of the howling dog."

MYTHOLOGICAL HISTORY.—Virgo was the Goddess Astræa. According to the poets, the early history of man was the golden age. It was a time of innocence and truth. The gods dwelt among men, and perpetual spring delighted the earth. Next, came the silver age, less tranquil and serene, but still the gods lingered and happiness prevailed. Then followed the brazen and iron ages, when wickedness reigned supreme. The earth was wet with slaughter. The gods left the abodes of men, one by one, Astræa alone remaining; until finally she too, last of all the immortals, bade the earth farewell. Jupiter thereupon placed her among the constellations.

Hydra is a long, straggling serpent, having its head near Procyon and extending its tail beyond Virgo, a total distance of more than 100°.

THE PRINCIPAL STAR is Cor Hydræ, of the second magnitude. It is a lone star, and may be easily found by a line drawn from γ Leonis through Regulus, and continued about 23°. The head is marked by a rhomboidal figure of four stars of the fourth magnitude lying near Procyon. Several little triangles may be formed of them and other small stars lying near. The *Crater*, or Cup, is a beautiful and very striking semicircle of six stars of the fourth magnitude directly south of θ Leonis. *Corvus* (β, ε, γ, δ), the raven, lies 15° east of the Cup. ε Corvi is in the equinoctial colure.

MYTHOLOGICAL HISTORY.—Hydra was a fearful serpent which in ancient times infested the lake Lerna. Its destruction constituted one of the twelve labors of Hercules. The Crow was formerly white, it is said, but was changed to its raven tint on account of its proneness to tale-bearing.

Canes Venatici, the hunting dogs. This constellation contains the bright star, Cor Caroli (α), which is found by a line passing from Benetnasch (η) through Berenice's Hair to Denebola (β).

Berenice's Hair is a beautiful cluster midway between Cor Caroli and Denebola. Near by is a single bright star of the fourth magnitude.

MYTHOLOGICAL HISTORY.—Berenice was the wife of Ptolemy. Her husband going upon a dangerous expedition, she promised to consecrate her beautiful tresses to Venus if he should return in safety. Soon after the fulfilment of this vow the hair disappeared from the temple where it had been deposited. Berenice being much disquieted at this loss, Conon, the astronomer, announced that the locks had been transferred to the heavens, in proof of which he pointed out this cluster of hitherto unnamed stars. All parties were satisfied with this happy termination of the difficulty.

Boötes, the bear-driver, is represented as a hunts-man grasping a club in his right hand, while in his

(Map No. 5)—Fig. 91.

left he holds by the leash his two greyhounds (*Canes Venatici*), with which he is pursuing the Great Bear continually around the north pole.

PRINCIPAL STARS.—Arcturus (Job, ix, 9), a mag-nificent star of the first magnitude, is in the left knee. It forms a triangle with Denebola and Spica, and also one with Denebola and Cor Caroli. It travels in its orbit fifty-five miles per second, or three times as fast as the earth (p. 205).   Its light reaches the earth in twenty-five years. Mirach (ϵ) lies in the girdle, δ in the right shoulder, Alkaturops (μ) in the club, β in the head, and Seginus (γ) in the left shoulder. Seginus forms with Cor Caroli and Arcturus a triangle, right-angled at Seginus. Three small stars in the left hand of Boötes lie near Benetnasch.

MYTHOLOGICAL HISTORY.—Boötes is supposed to have been Arcas, the son of Callisto.   (See Ursa Major.)

Hercules is represented as a warrior clad in the skin of the Nemæan lion, holding a club in his right hand and the dog Cerberus in his left. His foot is near the head of Draco, while his head lies 38° south and his club reaches 10 degrees beyond.

THE PRINCIPAL STAR is Ras Algethi (a Herculis). This forms a triangle with β and δ. A peculiar figure of four stars (π, η, ζ, ϵ), north of these, marks the body. (See Maps, Nos. 5, 6, and 7.) The left knee is pointed out by θ, and the left foot by y.

MYTHOLOGICAL HISTORY.—This constellation immortalizes the name of one of the greatest heroes of antiquity.   Hercules was the son of Jupiter and Alcmena.   While he was yet lying in his cradle, Juno, in her jealousy, sent two serpents to destroy him.   The precocious infant, however, strangled them with his hands.   By the cunning artifice of Juno, Hercules was made subject to Eurystheus, his elder half-brother, and compelled to perform all his commands.   Eurystheus enjoined upon him a series of the most difficult and dangerous enterprises that could be conceived, which have been termed the "Twelve Labors of Hercules."   Having completed these tasks, he afterward achieved others equally celebrated.   Near the close of his life he killed the centaur Nessus.   The dying monster charged Dejanira, the wife of Hercules, to preserve a portion of his blood as a charm to use in case the love of her husband should ever fail her.   In time, Dejanira thought she needed the potion, and Hercules having sent for a white robe to wear at a sacrifice, she steeped the garment in the blood of Nessus.   No sooner had Hercules put on the fatal robe than the venom stung his bones and boiled through his veins.   He attempted to tear it off, but in vain. It stuck to his flesh, and tore off great pieces of his body.   The hero, finding he must die, ascended Mount Œta, where he erected a funeral pyre, spread out the skin of the Nemæan lion, and laid himself down upon it. Philoctetes applied the torch.   With perfect serenity of countenance Hercules awaited approaching death—

> "Till the god, the earthly part forsaken,
> From the man in flames asunder taken,
> Drank the heavenly ether's purer breath.
> Joyous in the new unwonted lightness
> Soared he upward to celestial brightness,
> Earth's dark, heavy burden lost in death."—SCHILLER.

Corona consists of six stars arranged in a semi-circular form.   The brightest of these is Alphecca. This makes a triangle, with Mirach (ε) and δ in Boötes. It forms a similar figure with Mirach and Arcturus.

Serpentarius, or *Ophiuchus,* the serpent-bearer, is represented under the figure of a man grasping in both hands a prodigious serpent, which is writhing in his grasp.

PRINCIPAL STARS.—Ras Alhague (α), in the head, is of the second magnitude.   It is about 5° from Ras Algethi.   They form a pair of stars conspicuous like the pairs in Gemini, Canis Minor, Canis Major, etc.; β marks the right shoulder, and κ the left.   There is a small cluster near β, called *Taurus Poniatowskii.* An irregular square of four stars, near γ Herculis, denotes the head of the serpent.

MYTHOLOGICAL HISTORY.—This constellation perpetuates the memory of Æsculapius, the father of medicine.   He was so skilful that he restored several persons to life; whereupon Pluto complained to Jupiter that his kingdom was in danger of being depopulated. Therefore Jupiter struck him with a thunderbolt, but afterward placed him among the constellations. Serpents were sacred to Æsculapius, because of the superstitious idea that they have the power of renewing their youth by changing their skin.

Libra represents the scales of Astræa (Virgo), the goddess of justice. It may be recognized by the

(Map No. 6)—Fig. 92.

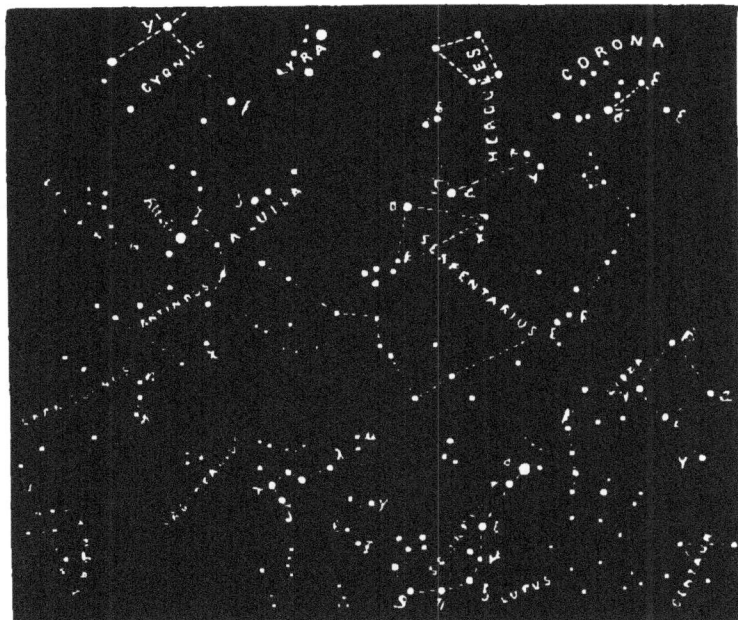

quadrilateral figure formed by its four principal stars.

Scorpio is represented under the figure of a huge scorpion, stretching through 25°. It is a most interesting constellation.

PRINCIPAL STARS.—Antares (*a*) is a fiery red star of the first magnitude. It marks the heart of the Scorpion. The head is indicated by several stars, the most prominent of which is β, arranged in a line slightly curved. The tail may be easily traced by a series of stars which winds around through the Milky Way in a beautiful manner.*

MYTHOLOGICAL HISTORY.—This is the scorpion that sprang out of the earth at the command of Juno, and stung Orion. Scorpio and Orion are so placed among the constellations that they never appear in the heavens together.

Sagittarius, the archer, is represented as a centaur with his bow bent, as if about to let fly an arrow at Scorpio.

PRINCIPAL STARS.—A row of stars from μ to β marks the bow: another from γ eastward points out the arrow and the right arm drawn back in bending the bow. North of τ, two stars of the fourth magnitude denote the head of the centaur. The *Milk Dipper*, so called because the handle lies in the Milky Way, is a very striking figure.

MYTHOLOGICAL HISTORY.—This constellation is named in honor of Chiron, one of the centaurs. These monsters were represented as men from the head to the loins, while the remainder of the body was that of a horse —the ancients having so high an opinion of that animal that the union was not considered in the least degrading.

Chiron was renowned for his skill in music, medicine, and prophecy. The most distinguished heroes of mythology were among his pupils. He taught Æsculapius physic; Apollo, music; and Hercules, astronomy.

Capricornus contains no very conspicuous stars. The *Southern Fish* (No. 6) has one star of the first magnitude, Fomalhaut (α, No. 7), which on a clear summer evening may be seen in the south, midway to the zenith. *Antinous and the Eagle* is a double

* Antares (*anti*, like ; *Arēs*, Mars) was so named because it rivalled Mars in brightness and color.

constellation. It contains a beautiful star of the first magnitude, Altair. This is conspicuous, as being the center one in the row of three bright stars. A similar row denotes the tail of the eagle; the first star of which is named ζ, and the last star lies in Cerberus. The *Dolphin* contains a pretty cluster in the form of a diamond. It is sometimes called *Job's Coffin*.

(Map No. 7)—Fig. 93.

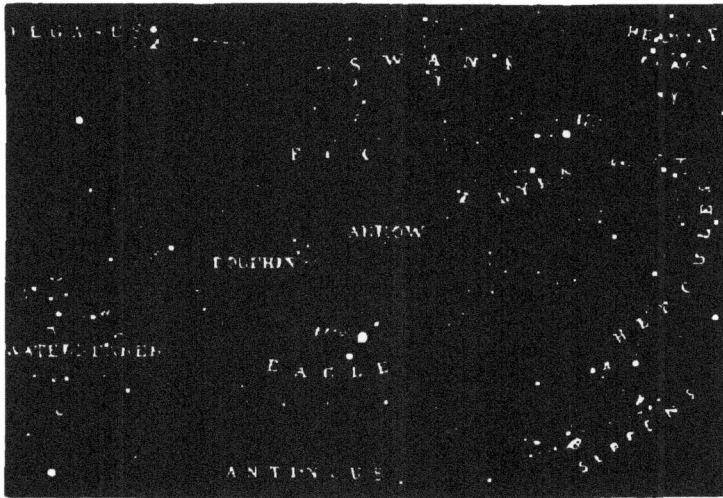

Cygnus, the swan, is a remarkable group of stars. the principal ones being so arranged as to form a large and beautiful cross. The upright piece lies along the Milky Way. It is composed of four stars, three of which, Deneb (α), γ, and β, are bright, while the fourth is a variable star. No. 61, a minute star, scarcely visible to the naked eye, is noted as being the nearest to the earth of any of the fixed stars in the northern hemisphere (p. 241).

Lyra, the harp, contains one brilliant blue star, Vega (p. 217). Close by it is a parallelogram of four smaller stars, by which it may be easily recognized.

MYTHOLOGICAL HISTORY.—This is the celestial lyre upon which Orpheus discoursed such ravishing music that wild beasts forgot their fierceness and gathered about him to listen, while the rivers ceased to flow, and the very rocks and trees stood entranced.

III. Southern Constellations.—We now imagine ourselves viewing the stars visible to a person far

(Map No. 8)—Fig. 94.

south of the equator. The constellations are reversed with reference to the horizon. The two stars which, in the northern hemisphere, compose the base of the parallelogram in Orion, form here the upper side. Sirius is above Orion. All the northern circumpolar constellations are hidden from view. At the southern pole there is no conspicuous star, but the richness and number of the neighboring stars compensate this deficiency, and give to the heavens an incomparable splendor. Here is the magnificent constellation Argo, in which we find Canopus, looked upon anciently as next to Sirius in brilliancy : η, a variable star, now surpasses it in brightness.

Nearly at the height of the south pole, blazes the *Southern Cross;* below is the *Centaur,* containing two stars of the first magnitude and five of the second ; and above is Hydrus, where shines Achernar, another beautiful star of the first magnitude.

(Map No. 9)—Fig. 95.

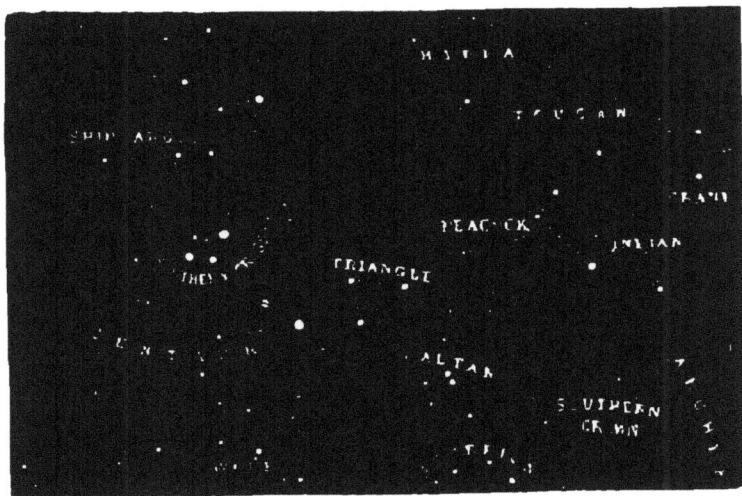

III.   DOUBLE STARS, COLORED STARS, NEBULÆ, ETC.

1. Double Stars.—To the naked eye, all the stars appear single. With the telescope, over 10,000 have been found to be double. Thus, Polaris consists of two stars about 18″ apart; Rigel has a companion about 10″ from it; and Sirius, one distant 7″.   A good opera-glass will separate ε Lyræ into two components.

In case two stars happen to lie in the same straight line from us, though at immense distances from each other, their light will blend. They will be seen by the naked eye as a single star, and by the telescope as a double star. They are called *optical double* stars. Many, however, of the double stars have been found to be *physically* connected. Each double star of this class forms a binary system of two suns revolving in an elliptical orbit about their common center of gravity, like the planets in the solar system, in accordance with Newton's law of gravitation. In a few instances, there are combinations of *triple*, *quadruple*, and even *septuple* stars. Thus ε Lyræ is a *double-double* star, while θ Orionis is a system of six suns. The components of a double star commonly differ in brightness; so that frequently the fainter one is nearly lost in the brilliancy of its companion sun.

THE PERIODS of some systems have been ascertained. Thus, ξ Ursæ Majoris is a double star, and the two stars of which it is composed have performed an entire revolution about each other since they were found to be connected. There are only eleven binary stars now known whose periods are less than a century, while the others have periods which seem to extend, in some cases, beyond a thousand years.

ORBITS.—It is not possible to estimate the dimensions of the orbits of the double stars, until their distances from us are definitely known. " Taking the estimated distance of 61 Cygni (550,000 times the sun's distance from the earth)* as a basis, the companions of that system cannot cultivate a very intimate acquaintance, since they must be over a billion miles apart. From these data, astronomers have attempted even to calculate the mass of some of the double stars. 61 Cygni, although scarcely visible to the naked eye, and known to be the second nearest to us of any of the fixed stars, is estimated to weigh one-third as much as our sun." (See p. 308.)

II. Colored Stars.—We have already noticed that the stars are of various colors.† Sirius is white; Antares, red; and Capella, yellow ; while Lyra has a blue tint, and Castor has a green one. In the pure transparent atmosphere of tropical regions, the colors are far more brilliant. There, oftentimes, the nocturnal sky is a blaze of jewels,—the stars glittering with the green of the emerald, the blue of the amethyst, and the red of the topaz.

In the double and multiple stars, every color is presented in all its richness and beauty ; while there are also combinations of colors complementary to each other. Here is a green star with a blood-red companion ; here an orange and a blue sun ; there a yellow and a purple one. The triple star γ Andromedæ is formed of an orange-red sun and two others of an emerald green.

Every tint that blooms in the flowers of summer,

---

* Recent measurements of this star seem to indicate its probable distance from the sun to be 400,000 radii of the earth's orbit.

† The theory has been advanced that the color indicates the intensity of the heat of the star. A white star is therefore hotter than a red star ; and a blue, than a yellow one

flames out in the stars at night. "The rainbow flowers of the footstool and the starry flowers of the throne," proclaim their common Author ; while rainbow, flower, and star alike evince the same Divine love of the beautiful.

We can hardly conceive the effects produced in a system having colored suns. Take a planet revolving about ψ Cassiopeiæ for instance. This is illuminated by a red, a blue, and a green sun. Sometimes, by the succession of these suns, a cheerful green day would present a charming relief to a fiery red one ; and that might be still further subdued by a gentle blue one. The odd contrast of color and the vicissitudes of extreme heat and cold that obtain on such a world, present a picture which our fancy can sketch better than words can paint.

The colors of the stars change. Sirius was anciently red. It is now unmistakably white. There are two double stars which were described by Herschel as white ; each is now composed of a golden-yellow and a greenish star.

III. The Variable Stars have periodic changes of brilliancy. The following are most conspicuous :

ALGOL, in the head of Medusa, is a star of the second magnitude for about two and a half days, when it suddenly decreases, and in three-and-a-half hours descends to the fourth magnitude. It then rekindles, and in three-and-a-half hours is again as brilliant as ever.

MIRA, the wonderful, a star in the Whale, has a period of eleven months. It is ordinarily of the second magnitude for about fifteen days. It then decreases for three months, until it becomes invisible to the naked eye. This period of darkness lasts five months; it then rebrightens for three months, until it regains its former lustre. Occasionally, however, it fails to brighten at all beyond the fourth magnitude, while on one occasion its light was almost equal to that of Aldebaran. Sometimes no perceptible change takes place for a month; then again, there is a sensible alteration in a few days.

THE REASON OF THIS VARIABILITY is not understood. It has been suggested, in the case of Mira, that it may be a globe rotating on its axis, and that different portions of its surface, illuminated to different degrees of intensity, are thus presented to us. Others have conceived that there may be satellites revolving about these suns, and that when their dark bodies interpose between the stars and our earth, they eclipse the light wholly or in part.

IV. The Temporary Stars suddenly blaze out in the heavens, and then gradually fade away. The most celebrated one burst forth in Cassiopeia, in the year 1572. Tycho Brahe says: "One night as I was examining the celestial vault, I saw with unspeakable astonishment a star of extraordinary brightness in Cassiopeia. Struck with surprise, I could scarcely believe my eyes. To convince myself that there was no illusion, I called the workmen of my laboratory and the passers-by, and asked them if they saw the star which had so suddenly made its appearance. It could be compared only with Venus at her quadrature, being seen distinctly at midday." Its color was at first white, then yellow, and finally red. Its brightness decreased gradually until the spring of 1574, when the star disappeared from view and has not since been seen. As two brilliant stars had previously appeared in Cassiopeia, at intervals of about three centuries, they have been thought, by some, to be identical, and that it is only a variable star of long period.

Since this discovery by Tycho Brahe, numerous instances are recorded of stars which have suddenly burst forth, and have then either faded out entirely, or remain as faint telescopic objects. In the latter case, they are termed *New stars*. One of this kind appeared in Corona Borealis, in 1866. At first it was of the second magnitude, but in a week changed to the fourth, and in a month diminished to the ninth. Strangely, too, some stars have disappeared from the heavens, and are styled *Lost stars*. The changes which are thus constantly taking place are calculated to make the term "eternal stars" seem a very indefinite phrase.

EXPLANATION.—These phenomena are as yet little understood. A rotation about an axis would fail to explain the changes in color. Some think that these stars revolve in enormous orbits of such eccentricity that at their most distant points they fade out of sight. Arago has shown, in reply to this, that for a star to decrease in brightness from the first magnitude to the second by moving directly from us, even with the velocity of light, would require six years. As we have just seen, the star of 1866 underwent this change in brilliancy in a week.

The mind cannot help wondering if they are not

instances of enormous conflagrations in which a world is overwhelmed in ruin ! The investigations of spectrum analysis indicate that the star of 1866 consisted of *burning hydrogen gas*. We can suppose that the gas was evolved by some convulsion, and, taking fire, wrapped the entire globe in flames. This need not involve the idea of destruction, but only a change of form. A dark star may thus become luminous, or a bright one may be extinguished.*

5. **The Star Clusters** are groups of stars so massed together as to present a hazy, cloud-like appearance. Several of them have been already named,—the Pleiades, the Beehive in Cancer, Berenice's Hair, the Hyades, and the group in the sword-handle of Perseus. The principal stars of which they are composed can generally be distinguished by the naked eye, although by the use of a small opera or spy-glass the number is increased.

In the southern sky, there are clusters still more remarkable. In the Cross, is a group of 110 stars of various colors, red, blue, and green, so that looking on it, says Herschel, is "like gazing into a casket of precious gems." A cluster in Toucan is compact in the center, where it is of an orange-red color ; the exterior is composed of pure white stars, making a border of exquisite contrast.

It is generally conceded that there is some close

---

* The process of apparent creation and destruction is going on in the heavens immediately before the eye of the astronomer. New stars flash light, old stars are lost, worlds burst into flame, and their glowing embers fade into darkness. Are they re-created into new worlds? We know not. We only perceive that the same Almighty power which fitted up this earth for our home is yet at work among the worlds about us, and we are thus witnesses of His eternal presence.

Fig. 96.

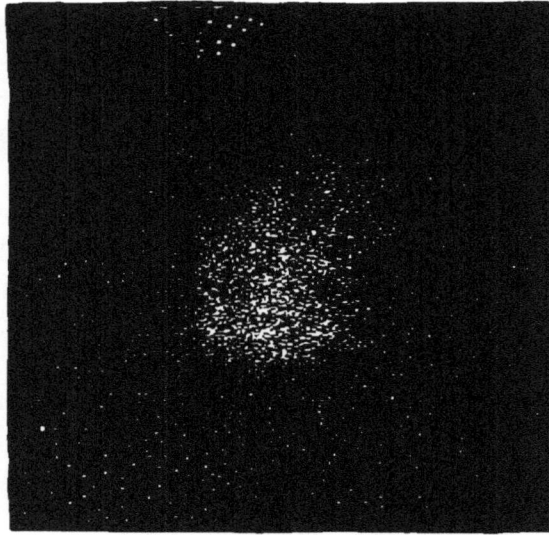

Star-Cluster in Toucan.

physical relation existing between the stars composing such an "archipelago of worlds," but its nature
is a mystery. They seem generally crowded together
toward the center, blending into a continuous blaze
of light. Yet, although they appear so densely compacted, it is probable that, if we could change our
stand-point and penetrate one of these groups of
suns, we should find it, on our approach, opening up
and spreading out before us, until, in the midst, the
suns would shine down upon us from the heavens as
the stars do in our own sky.

6. Nebulæ are faint, misty objects, like specks of
luminous clouds. A few are visible to the naked eye,
but the telescope reveals thousands. They differ
from clusters in not being resolvable into stars when
viewed through the largest telescopes. With the con-

stant improvement made in these instruments, however, many so-called nebulæ have been resolved, and thus the number of clusters has been increased, while new nebulæ have been discovered.

Until of late, it was thought that all nebulæ were simply groups of stars, which would be ultimately discerned in the more powerful telescopes yet to be made. Spectrum analysis shows, however, that many of these luminous clouds are gaseous, and are not composed of stars.

Since all the nebulæ maintain the same position with respect to the stars, their distance must be inconceivably great, and, in order to be visible to us, their magnitude must be proportionately vast. They are most abundant at the two poles of the Milky Way, but are more uniformly distributed over the heavens lying near the south pole.

It is now generally believed that nebulæ constitute the material for making stars,—are, in fact, *sun-germs;* that all stars originally existed as nebulæ; and that every nebula will, in time, be changed into stars.

Nebulæ are divided, according to their form, into six classes—*elliptic, annular, spiral, planetary, irregular nebulæ,* and *nebulous stars.*†

THE ELLIPTIC or merely oval nebulæ are the most abundant. Under this head is classed the *Great Nebula in Andromeda,* which was discovered over

---

† This division of the nebulæ is purely arbitrary, and used only to introduce some order of arrangement. The shape of the nebulæ changes with the power of the telescope through which they are seen. Thus the Great Nebula in Andromeda, as resolved by Bond, is no longer oval, but irregular in form. The Ring-Nebula of Lyra, seen through the large telescope of to-day, is egg-shaped; while the Dumb-bell Nebula assumes the outline of a chemical retort.

a thousand years ago, and is visible to the naked eye.   Prof. Bond, of the Cambridge Observatory,

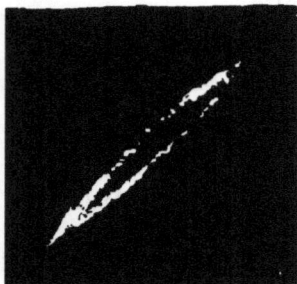

Fig. 97.

Nebula in Andromeda.

has partly resolved it into stars, of which he has counted 1500, although its nebulous appearance was still retained. Through the telescope, it is one of the most glorious objects in the heavens. "If we suppose this nebula to be one continuous bed of stars of different sizes for its entire extent, it must comprise the enormous number of 30,000,000."

The distance of such nebulæ from the earth passes our comprehension.   Some astronomers have estimated that a ray of light would require 800,000 years to span the gulf that intervenes. Imagination wearies itself in the attempt to understand these figures. They teach us something of the limitless expanse of that space in which God is working the mysterious problem of creation.

Fig. 93.

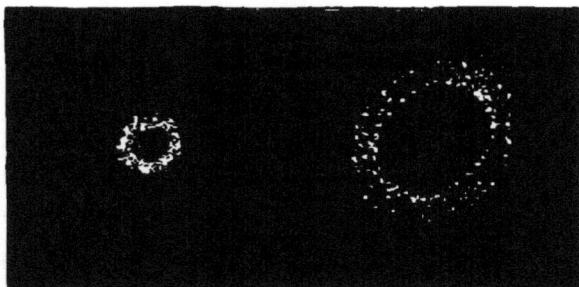

Nebula in Lyra.

THE ANNULAR NEBULÆ have the form of a ring. There are four of these "ring universes." In the cut

is a representation of one in Lyra,—first, as seen by
Herschel, having in the center a nebulous film like a

Fig. 99.

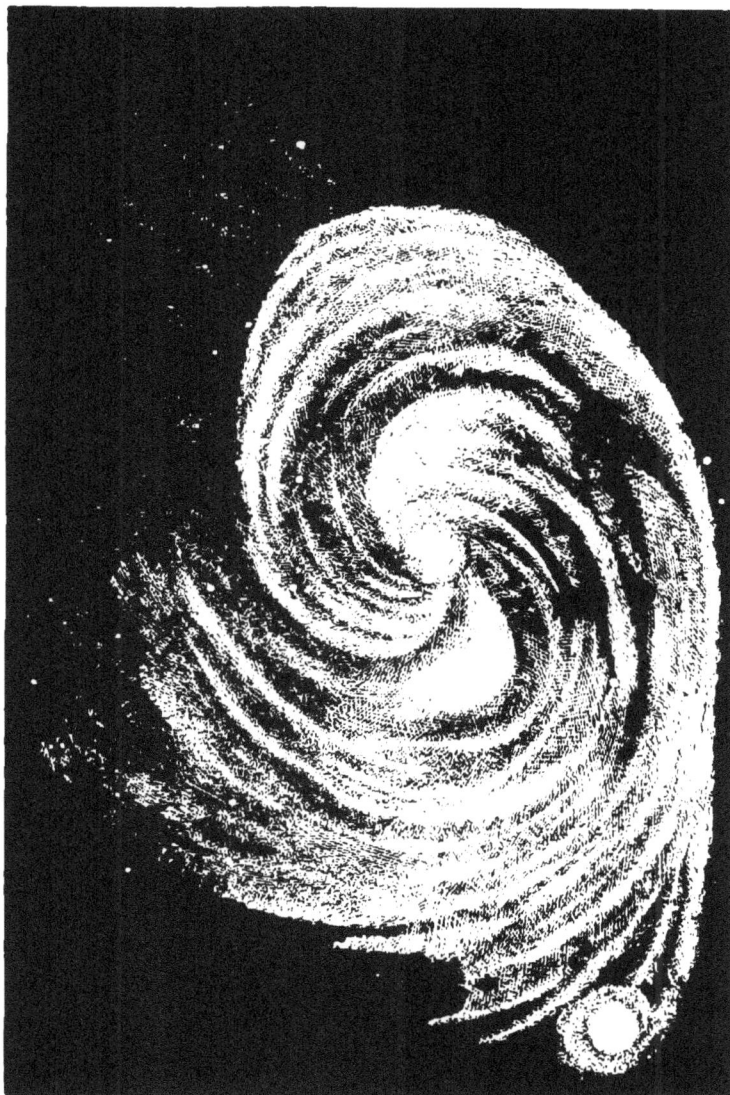

Spiral Cluster in Canes Venatici.

"bit of gauze stretched over a hoop;" second, as shown in Lord Rosse's telescope (p. II), which resolves the filmy parts of the nebula into minute stars, and reveals a fringe of stars along the edge.

THE SPIRAL or WHIRLPOOL NEBULÆ are exceedingly curious. The most remarkable one is in Canes Venatici. It consists of brilliant spirals sweeping outward from a central nucleus, and all overspread with a multitude of stars.* One is lost in attempting to imagine the distance of such a mass, and the forces which produce such a "tremendous hurricane of matter—perhaps of suns."

PLANETARY NEBULÆ, by their circular form and pale, uniform light, resemble the disks of the distant

Fig. 100.

planets of our system. Their edges are generally well defined, though sometimes slightly furred. There is one in Ursa Major, which, if located at the distance of 61 Cygni, would "fill a space equal to seven times the entire orbit of Neptune."

Planetary Nebula.

IRREGULAR NEBULÆ are those which have no definite form. Many present the irregularities of clouds torn by the tempest. Some of the likenesses which may be traced are strangely fantastic : for example, the Dumb-bell Nebula, in the constellation Vulpecula, and the Crab Nebula, near the southern horn of Taurus. There is also one known

* Columbus discovered a new continent, and so immortalized his name ; what shall we say of the astronomer who discovers a system of worlds?

as the Great Nebula in the Sword-handle of Orion, which bears a faint resemblance to the wings of a bird.

NEBULOUS STARS are so called because they are enveloped by a faint nebula, usually of a circular form. The star is generally seen at the center, although some nebulæ surround two stars,

Fig. 101.

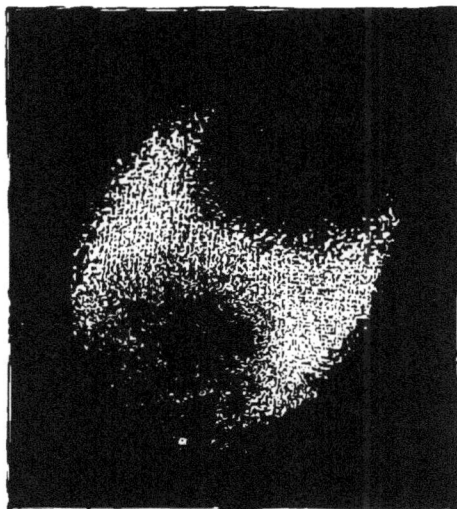

Dumb-bell Nebula.

having one in each focus. It is thought that these may be suns possessing immense atmospheres, which are rendered visible somewhat as that of our sun is in the zodiacal light; and that in like manner our sun may present to other worlds the appearance of a nebulous star.*

VARIABLE NEBULÆ.—Certain changes take place

* Nothing in all nature is more suggestive of the magnificence and immensity of Creation, than are the nebulous star-clusters, many of which are at such an inconceiv- able distance, that the most powerful telescopes show them only as a confused mass of light. A casual observer,—even though when led by scientific analogy to resolve each little patch of star-dust into a host of separate suns, and to provide each sun with a retinue of inhabited planets,—might think of them as little colonies of suns, set on the very outskirts of world-creation, and moving in such close proximity, that the peoples of the various worlds might communicate with one another. Yet, were he transported to some planet whirling about one of those far-off star-suns,—a multitude of which blend as a single point of light to our human eyes,—he would see the other suns only as fixed stars in the firmament above him ; and though many of them might surpass in splendor the glory of our own Sirius, yet all would still remain at such an immense distance as to baffle the research of the most powerful telescopic instruments. Thus, too, he would probably find each planet revolving at such a distance from its sister planets, as to render the certain knowledge of other inhabited worlds as elusive there as here.

Fig. 102.

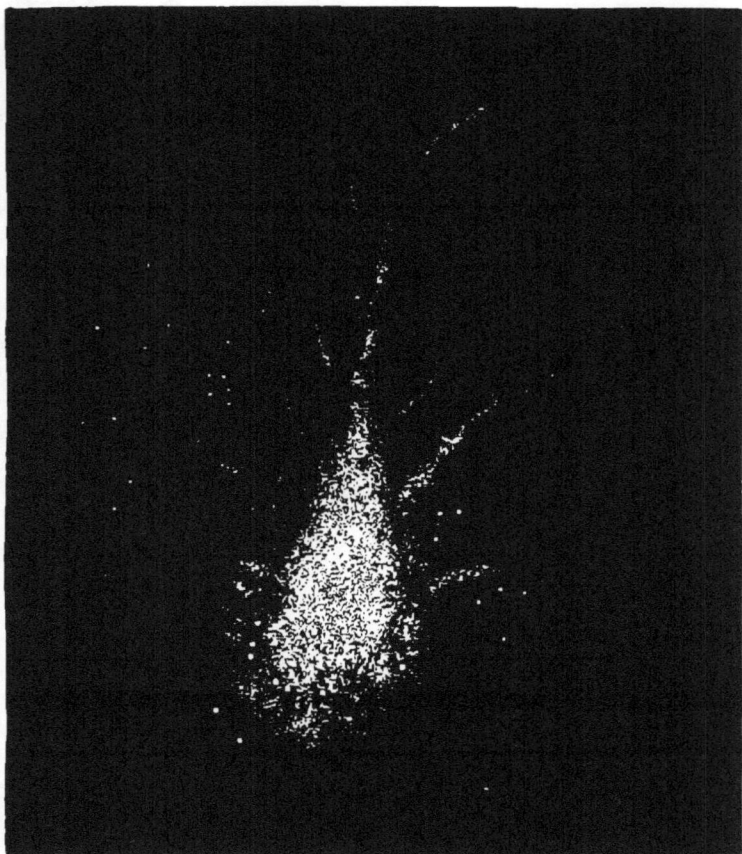

Crab Nebula.

among the nebulæ which can be accounted for only
under the supposition that they, like some of the
stars, are *variable.* Mr. Hind tells us of a nebula in
Taurus which, in 1852, was distinctly visible with a
small telescope, but, in 1862, had vanished entirely out
of the reach of a powerful instrument. The Great
Nebula in Argo, when observed by Herschel in 1838,

had in the center a vacant space containing a star of the first magnitude, enshrouded by nebulous matter. In 1863, the nebulous matter had disappeared, and the star was only of the sixth magnitude. These facts as yet defy explanation. They illustrate the vast and wonderful changes constantly taking place in the heavens.

DOUBLE NEBULÆ.—There seems to be a physical connection existing between some of the nebulæ, similar to that already noticed in respect to certain stars. In the case of the latter, this inter-relation has been proved, since, even at their distances, their movements can yet be traced in the lapse of years. "But, owing to the almost infinite depths in the abyss of the heavens at which these nebulæ exist, thousands of years, perhaps thousands of centuries, would be necessary to reveal any movement."— (Guillemin.)

7. **Magellanic Clouds.**—Not far from the southern pole of the heavens, there are two cloud-like masses, distinctly visible to the naked eye, known to navigators as Cape Clouds. Sir John Herschel describes them as consisting of swarms of stars, clusters, and nebulæ, seemingly grouped together in the wildest confusion. In the larger, he found 582 single stars, 46 clusters, and 281 nebulæ.

8. **The Milky Way.**—Via Lactea, or the Galaxy, is a luminous, cloud-like band that stretches across the heavens in a great circle. It is inclined to the celestial equator about 63°. This stream of suns is divided into two branches from α Centauri to Cygnus. To the naked eye, it presents merely a diffused

light; but with a large telescope it is found to consist of myriads of stars densely crowded together.*

These stars are not uniformly distributed through the entire extent. In some regions, within the space of a single square degree we can discern as many as can be seen with the naked eye in the entire heavens. In other parts, there are broad, open spaces. A remarkable instance of this occurs nears the Southern Cross. There is a dark, pear-shaped vacancy, with a single bright star at the center, glittering on the blue background of the sky. In viewing it, one is said to be impressed with the idea that he is looking through an opening into the starless depths beyond the Milky Way.

The northern galactic pole is situated near Coma Berenices, and the southern in Cetus. Advancing from either pole toward the Milky Way, the number of stars increases, at first slowly and then more rapidly, until the proportion at the galaxy itself is thirty-fold.

Fig. 103.

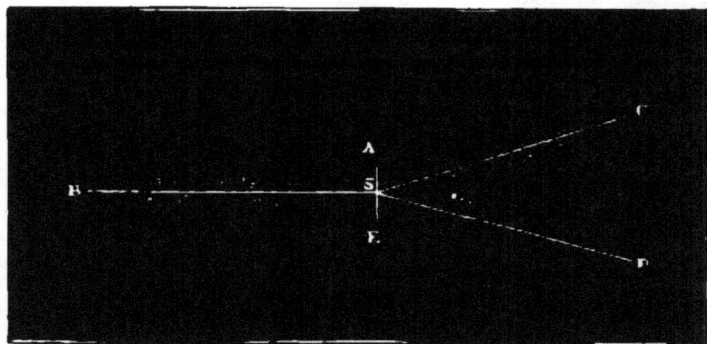

Herschel's Theory of the Stellar System.

* Herschel states that 258,000 stars once passed across the field of his great reflector in 41 minutes. With the powerful instruments now making, it is probable that many more could be seen.

HERSCHEL'S THEORY.*—Sir W. Herschel has conjectured that the stars are not indifferently scattered through space, but are collected in a stratum something like that shown in the cut, and that our sun occupies a place at S, near where the stream branches, A and E being the galactic poles. It is evident that, to an eye viewing the stratum of stars in the direction SB, SC, or SD, they would seem much denser than in the direction SA or SE. Thus are we to think of our own sun as a star of the second or third magnitude, and of our little solar system as plunged far into the midst of this vortex of worlds, a mere atom along that

" Broad and ample road
Whose dust is gold and pavement stars."

9. The Nebular Hypothesis is a theory advanced by Laplace, to show how the solar system may have been formed.† As since modified, its outlines are as follows: In the "beginning," all the matter which now composes the sun, and the various planets with their moons, was in a gaseous and highly heated state. It filled the space at present occupied by the system, and extended far beyond the orbit of Neptune. In other words, the solar system was simply an immense nebula. The heat, which is the repel-

---

* Other theories have been advanced by astronomers, but we are as yet ignorant of the real structure of the universe outside of our own system.

† We should remember that this theory aims to tell only the way in which our system was developed. The parent nebula must have contained a potential energy equal to all the manifestations of force since made in the entire system. Nothing could be developed from a mass of nebulous matter the germs of which had not been put in it originally by the Creator. The analogies of nature all go to show that the Creator's plan is, in general, not to produce any object in a perfect and matured state; but rather, by a gradual growth, to unfold its full form and function.

lant force, overcame the attraction of gravitation. Gradually the mass cooled by radiation. As centuries passed, the repellant force becoming weaker, the attractive force drew the matter and condensed it toward one or more centers. The nebula then presented the appearance of a nebulous star—a nucleus enveloped by a gaseous atmosphere.

According to a well-known law in physics, seen in every-day life, wherever matter seeks a center— as in a whirlpool, in a whirlwind, or even in water poured through a funnel—a rotary motion was established. As the rotary motion of the nebula increased, the centrifugal force finally overcame at the exterior the attraction of gravitation. A ring of condensed vapor was then left behind.* Centuries elapsed, and again, under the same conditions, a second ring was detached. Thus, one by one, concentric rings were separated from the parent nebula, all revolving in the same plane and in the same direction. These different rings, becoming gradually consolidated, formed the planets. Generally, however, in this process, while still in the vaporous state and slowly condensing, the rings themselves detached other rings that were in turn consolidated into satellites.

In the case of Saturn, several of these secondary rings did not condense into globes, but still remain as rings which revolve about the planet.† Mitchell

---

* A considerable modification of the Nebular Hypothesis is possible, leaving its general idea, however, intact. It is now generally conceded that the several planets were not "thrown off," but merely *detached* and left behind. Proctor thinks that the solar system is the result of *meteoric aggregation* as well as *gaseous condensation*: the planets in their infancy being so large, gathered immense quantities of meteoroids, then more abundant than now.

† In the case of the minor planets and the rings of Saturn, we may suppose that the

naïvely remarks, "Saturn's rings were left unfinished to show us how the world was made." The ring which formed the minor planets broke up into small fragments, none large enough to attract the rest and thus form a single globe.

The central mass of vapor finally condensed itself into the sun, which remains the largest member of the system. According to this theory, the sun may yet give off a few more planets, whose orbits will not exceed its present diameter. After a time, all its heat will be radiated into space, its fire will become extinct, and life on the planets will cease. We know not when this remote event may occur. We cannot fathom the purpose of God in creating and maintaining this system of worlds, nor can we foretell how soon it may complete its mission. We are assured, however,

> " That nothing walks with aimless feet,
> That not one life shall be destroyed,
> Or cast as rubbish to the void,
> When God hath made the pile complete."
>
> IN MEMORIAM.

---

rings were composed of matter uniformly distributed ; while in the case of the rings that consolidated into planets, there was a nucleus that attracted the rest of the matter to itself. It is possible that the rings of Saturn may yet break up and form new satellites for that planet. Indeed, some hold that one at least of the rings has thus been resolved into small meteorites. These may be attracted, and so picked up, one by one, in succession by the larger, until they form another moon, which will continue to revolve about the planet as the ring does now.—" The present state of the solar system is a living picture of the entire history of a single planet. From the sun's fire-mist, to ring-girt Saturn ; from Saturn, to storm-beaten Jupiter ; from Jupiter, to the sunny summer-time of our own planet ; from Earth, to autumn-browned Mars ; and from Mars, to the wintry silence and desolation of the dark gulches of the moon,—there is a series of stages that carries the thought back into the eternity long passed, as well as onward into the measureless depths of the future, and confers upon human intelligence a sort of exemption from the limitations of finite existence."—*Prof. Winchell.*

IV. CELESTIAL CHEMISTRY.

Spectrum Analysis.—The rainbow—that child of the sun and shower—is familiar to all. The brilliant band of colors, seen when the sunbeam is passed through a prism, is scarcely less beautiful. The ray of light containing the primary colors is here spread out fan-like, and each tint reveals itself. This variously-colored band is called a spectrum (plural, *spectra*). There are three different kinds of spectra—

1st. When the light of a solid or liquid body, as iron white-hot, is passed through a prism, the *spectrum is continuous*, and consists of a series of distinct colors, varying from red on one side to violet on the other.

2nd. If the light of a burning gas containing any volatilized substance be passed through a prism, the *spectrum is not continuous*, but is ornamented by bright-colored lines,—sodium giving two yellow lines ; strontium, a red one ; silver, two beautiful green ones. Each element produces a definite series which can be recognized as its test.

3rd. If a light of the first kind be passed through one of the second, the spectrum is crossed by *dark lines*. Thus, if the white light of an electric lamp be passed through a flame containing sodium, instead of the vivid yellow lines so characteristic of that metal, two black lines exactly occupy their place. *A gaseous flame absorbs the rays of the same color that it emits.* (See note, þ. 310.)

The Spectroscope.—This instrument consists of two small telescopes, with a prism mounted between their object-glasses (Fig. 106). The rays of light enter through a narrow slit at A, and are rendered parallel by the object-glass. They then pass through the prisms at C, are separated into the different colors, and, entering the second telescope at D, fall upon the

Fig. 106.

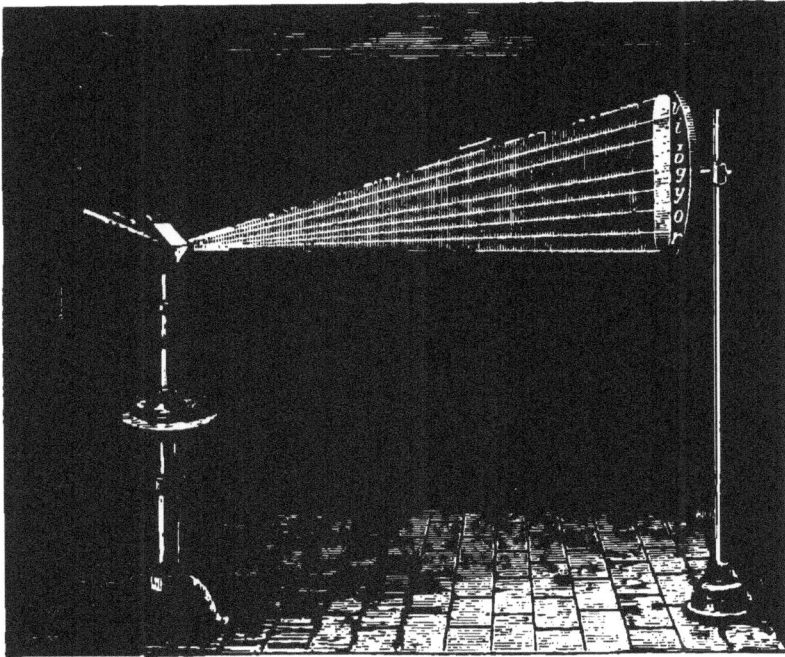

eye at B. A third telescope is sometimes attached, which contains a minutely-accurate scale for measuring the distances of the lines. In addition, a mirror may throw in at one side of the slit a ray of sunlight or starlight, and so we can compare the spectrum of the sunbeam with that of any flame we desire.

REVELATIONS OF THE SPECTROSCOPE CONCERNING
THE SUN.—The spectrum of the sunbeam is not continuous, but is crossed by a large number of dark
lines, called, from their discoverer, Fraunhofer's
lines. It is therefore concluded that the sun's light
is of the third class just named, and that it is pro-

Fig. 106.

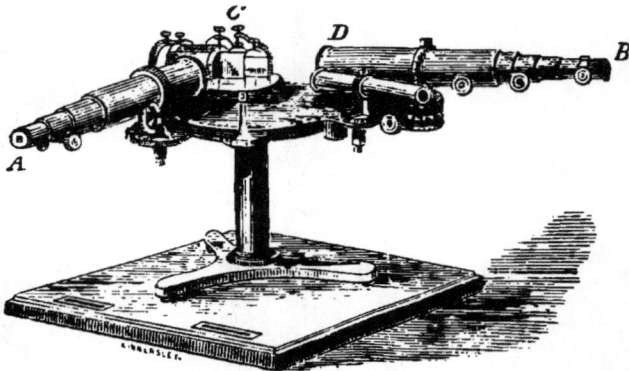

A Spectroscope.

duced by the vivid light of a highly heated body
shining through a flame full of volatilized substances.

But not only does spectrum analysis thus shed light
on the physical constitution of the sun, but these
lines are so distinctive, so marked and varied, that
the elements of which the sun is composed may be
discovered.* Thus, for example, iron gives a spectrum of some 450 lines, differing in intensity and

---

* The following twenty-two elements have been detected : sodium, calcium, barium,
magnesium, iron, chromium, nickel, cobalt, hydrogen, manganese, aluminium, titanium,
palladium, vanadium, molybdenum, strontium, lead, uranium, cerium, strontium,
cadmium, oxygen, and a probability of several more, such as carbon, silver, tin, etc.

relative length. These are bright when iron vapor is burning, and dark when white light is passed through such burning vapor. In the solar spectrum we have such a coincidence of dark lines, as to make the conclusion irresistible that iron is contained in the sun's atmosphere.*

Stars are Suns.—The same method of analysis has been applied to the stars. The spectra are marked by dark lines. Their constitution is therefore like our sun, and they also exhibit familiar elements. Betelgeuse, for example, contains many substances known to us, but, as is thought, no hydrogen. What a world that must be without water! We thus trace in the faintest star that trembles in the measureless depths of space, the elements that compose the common objects of our own life. We know that we are akin to nature everywhere,—a part of a system vast as the universe.

THE MOTION OF A STAR may be resolved into two components : one representing its motion at right angles, and the other its motion parallel to the line of vision. The former component can be determined by the telescope ; the latter is revealed by the spectro-

---

* Recent researches in spectroscopy present important problems. On elevating the temperature, it has been found that not only the lines of the spectrum of a substance vary, but new ones appear. Certain substances have apparently common lines. A molecule containing a few atoms gives a *line-spectrum ;* increase the number of atoms and it presents a *fluted-spectrum* (i. e., one composed of bands, each made up of lines, and having a sharp boundary on one side but fading away on the other) ; increase the number yet more, and it yields a *continuous spectrum.* New queries have therefore arisen in Solar Physics. How many atoms are there really in a specified molecule? What is the meaning of certain unfamiliar lines seen in the solar spectrum? Why do we not detect in the sun many of those substances that form so large a part of the earth's crust? Lockyer supposes that the so-called elements are really compounds whose molecules may be "dissociated" by intense heat, so that in the sun we see only the germs of our familiar chemical forms. Read Lockyer's "Spectrum Analysis," and Young's "The Sun."

scope.   If the star is moving towards us, the number
of vibrations producing any color will be increased,
and hence the dark lines corresponding to that color
in the spectrum will be pushed beyond its usual
place toward the violet end ; if going from us, the
number will be decreased, and the dark lines be
pushed toward the red end of the spectrum.*   The
amount of displacement once determined, the velocity of the star can be reckoned by means of well-
known laws of optics.

Spectra of Nebulæ.—Instead of being marked
with dark lines, as are the spectra of the stars, many
of the nebulæ exhibit bright lines.   Their spectra
are, therefore, of the 2nd kind.   This proves such
nebulæ to consist, not, like the stars, of an intensely-
heated nucleus shining through a luminous atmosphere, but of a glowing mass of gas.†   Out of 60
nebulæ examined by Mr. Huggins, 20 exhibited the
bright lines belonging to the gases, and all contained
nitrogen.

The Solar Flames, which were formerly seen only
during an eclipse, can now be examined by means of
the spectroscope, at any time.‡   The sun has thus

* The same result is produced in the case of sound. The whistle of an approaching
train sounds shriller than when it is receding. See Physics, p. 133.

† The Dumb-bell nebula is said to emit a light only about one twenty-thousandth
part that of a common wax-candle. If this matter be a " sun-germ," how immensely
must it become condensed before its rushlight glimmering can rival the dazzling brilliancy of even our own snn !

‡ " The red portion of the spectrum will stretch athwart the field of view like a scarlet
ribbon with a darkish band across it ; and in that band will appear the prominences,
like scarlet clouds, so like our own terrestrial clouds, indeed, in form and texture, that
the resemblance is quite startling.   One might almost think he was looking out
through a partly-opened door upon a sunset sky, except that there is no variety or contrast of color ; all the cloudlets are of the same pure scarlet hue. Along the edge of the
opening is seen the chromosphere, more brilliant than the clouds which rise from it or
float above it, and, for the most part, made up of minute tongues and filaments."—*Young.*

been found to be a sea of fire swept by the most violent storms.* Flames travel over its surface with a velocity of which we can form no conception ; "one jet shot out 80,000 miles and disappeared in ten minutes." Young describes a protuberance that reached the enormous height of 350,000 miles and then faded entirely away, all within two hours.

V. TIME.

Sidereal Time.—A sidereal day is the exact interval of time in which the earth rotates on its axis. It is found by marking two successive passages of a star across the meridian of any place. This is so absolutely uniform, that, as recent investigations seem to show, the length of the sidereal day has not varied more than $\frac{1}{80}$ of a second in 2,400 years, (note, p. 89).

The sidereal day is divided into twenty-four equal portions, which are called sidereal hours, and each of these hours into sixty portions, termed sidereal minutes, etc.

ASTRONOMICAL CLOCKS are regulated to keep sidereal time. The day commences when the vernal equinox is on the meridian. Therefore, the time by a sidereal clock does not point out the hour of the ordinary day. It indicates only how long it is since the vernal equinox crossed the meridian, and thus shows the right ascension of any star which may

---

* Such a storm "coming down upon us from the north would in 30 seconds after it had crossed the St. Lawrence be in the Gulf of Mexico, carrying with it the whole surface of the continent in a mass, not of ruin simply, but of glowing vapor, in which the vapors arising from the dissolution of the materials composing the cities of Boston, New York, and Chicago would be mixed in a single undistinguishable cloud."—*Newcomb.*

happen to be on the meridian at that moment. The hours of the clock are easily reduced to degrees (p. 28). The astronomer always reckons the hour of the day consecutively up to twenty-four.

Solar Time.—A solar day is the interval between two successive passages of the sun across the meridian of any place. If the earth were stationary in its orbit, the solar day would be of the same length as the sidereal; but, while the earth is turning around on its axis, it is going forward at the rate of 360° in a year, or about 1° per day. When the earth has made a complete rotation, it must therefore perform a part of another rotation through this additional degree, in order to bring the same meridian vertically under the sun.

One degree of diurnal rotation is equal to about four minutes of time. Hence the solar day is four minutes longer than the sidereal day. For the convenience of society, it is customary to call the solar day 24 hours long, and make the sidereal day only 23 hr. 56 min. 4 sec. in length, expressed in mean solar time. A sidereal day being shorter than a solar one, the sidereal hours, minutes, etc., are shorter than the solar; 24 hours of mean solar time being equal to 24 hr. 3 min. 56 sec. of sidereal time.

From what has been said, it follows that the earth makes 366 rotations around its axis in 365 solar days.

Mean Solar Time.—The solar days are of unequal length. To obviate this difficulty, astronomers suppose a *mean sun* moving through the equator of the heavens (which is a circle and not an ellipse) with a perfectly uniform motion. When this mean sun passes the meridian of any place, it is *mean noon;* and when the true sun is in the same position, it is *apparent noon.* This mean day is the average length of the solar days in the year. The clocks in common use are regulated to keep mean time.* When it is twelve by the clock, the sun may be either a little past or a little behind the meridian.

The difference between sun-time (apparent solar-time) and clock-time (mean time) is called the *"Equation of time."* This is the greatest about the first of November, when the sun is over sixteen and a quarter minutes in advance of the clock. The sun is the slowest about February 10th, when it is about fourteen and a half minutes behind mean time.

Mean and apparent time coincide four times in the year—namely, April 15th, June 14th, September 1st, and December 24th. On these days, the noon-mark on the sun-dial coincides with twelve o'clock.

The Sun-Dial.—The *apparent time* of the dial may be readily changed to mean time, by adding or subtracting the number of minutes given in the almanac for each day in the year, under the heading "sun slow" or "sun fast." A noon-mark is thus a very convenient method of regulating a timepiece.†

---

* In France, until 1816, apparent time was used ; and the confusion was so great, that Arago relates how the town clocks would differ thirty minutes in striking the same hour. As the time varied every day, no watchmaker could regulate a watch or clock to keep it.

† The following manner of obtaining one without a transit instrument may be useful. Select a level hard surface which is exposed to the sun from about 9 A. M. to 4 P. M. Upon this carefully describe, with compasses, a circle of eight or ten inches in diameter. Take a piece of heavy wire, six or eight inches in length, one end of which is sharpened. Drive this *perpendicularly* into the center of the circle, leaving it just high enough to allow the extreme end of its shadow to fall upon the circle about 9½ or 10 A. M. Mark this point, and also the place where the shadow touches the circle in the afternoon. Take a point half-way between the two, and, drawing a line from that to the center of the circle, it will be the *meridian line,* or noon-mark.

Why the Solar Days are of Unequal Length.—There are two reasons for this,—the unequal orbital motion of the earth, and the obliquity of the ecliptic. First: the orbit of the earth is an ellipse; and thus the apparent yearly motion of the sun along the ecliptic is variable. In perihelion, in January, the sun appears to move eastward daily 1° 1' 9".9; while at aphelion, in July, only 57' 11".5. As the earth in its diurnal motion rotates *uniformly* from west to east, and the sun passes eastward *irregularly*, this must produce a corresponding variation in the length of the solar day. The sun, therefore, comes to the meridian sometimes earlier and sometimes later than the mean noon, and they agree only at perihelion and aphelion.

Second: as we have just seen, the mean sun is supposed to move in a circle and not an ellipse. This would make the motion along the ecliptic uniform, but the obliquity of the ecliptic would still cause an irregularity in the length of the day. The mean

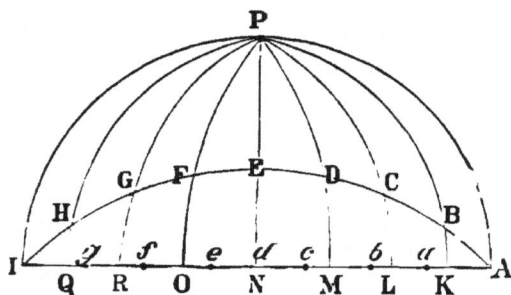

Fig. 107.

sun is therefore supposed to pass along the equinoctial, which is perpendicular to the earth's axis: while the ecliptic is inclined to it 23° 27'. Let A represent the vernal equinox; I, the autumnal; AEI, the ecliptic; AI, the equinoctial; PK, PL, PM, etc., meridians. Let the distances AB, BC, CD, etc., be equal arcs of the ecliptic, which are passed over by the sun in equal times. Next, on the equinoctial, mark off distances Aa, ab, bc, etc., equal to AB, BC, etc. These are equal arcs of right ascension, or hour-circles, through which the earth, rotating from west to east, passes in equal times. Now, meridians drawn through these divisions, would not agree with those drawn through equal divisions on the ecliptic. Hence, a sun moving along the ecliptic, which is inclined, would not make equal days, even though the ecliptic were a perfect circle.

Let us see how the mean and apparent solar days would compare. Let the real sun pass in its eastward course from A to B in a certain time ; the mean sun moving the same distance would reach the point a, since the latter travels on the base and the former the hypothenuse of a triangle. The earth, rotating from west to east, would cause the real sun to cross any meridian earlier than the mean sun ; hence, apparent time would be faster than clock-time. By holding the figure up above us toward the heavens, we can see how a westerly sun would cross the meridian earlier than an easterly one. Following the same reasoning, we can see that at the solstice, solar and mean time would agree ; while beyond that point the mean time would be faster.

The Civil Day is the mean solar day. It extends from midnight to midnight.* The method of divid-

* Until recently, very many nations terminated one day and commenced the next at sunset. Under this plan, 10 o'clock on one day would not mean the same as 10 o'clock on another day. The Puritans commenced the day at 6 P. M. The Babylonians, Persians, and Assyrians began the day at sunrise.

ing the day into two portions of twelve hours each, is said to have been adopted by Hipparchus, 150 years B. C., and is now in use over the civilized world. The astronomical method of reckoning the hours consecutively up to twenty-four is much more convenient, and is therefore coming into general favor. The names of the days are derived as follows:

1. Dies Solis.....Latin ....Sun's day.
2. Dies Lunæ....   "   ... Moon's day.
3. Tius daeg......Saxon....Tius's day.
4. Wodnes daeg... "   ....Woden's day.
5. Thurnes daeg.. "   ...Thor's day.
6. Friges daeg.....   "   ....Friga's day.
7. Dies Saturni....Latin ....Saturn's day.

The Year.—The *sidereal year* is the interval of a complete revolution of the earth about the sun, measured by a fixed star. It comprises 365 d., 6 hr., 9 min., 9.6 sec. of mean solar time. The *mean solar year* (tropical year) is the interval between two successive passages of the sun through the vernal equinox. It comprises 365 d., 5 hr., 48 min., 46.7 sec. If the equinoxes were stationary, there would be no difference between the sidereal and the tropical year. As the equinoxes retrograde along the ecliptic 50″ of space annually, the former is 20 min., 20 sec. longer.

The *anomalistic year* is the interval between two successive passages of the earth through its perihelion, which moves eastward about 11″.8 annually. It is 4 min., 40 sec. longer than the sidereal year.

The Ancient Year.—The ancients ascertained the length of the year by means of the *gnomon*. This was a perpendicular rod standing on a smooth plane on which was a meridian line. When the shadow cast on this line was the shortest, it indicated the summer solstice; and when it was the longest, the winter solstice. The number of days required for the sun to pass from one solstice back to it again determined the length of the year. This they found to be 365 days. As that is nearly six hours less than the true solar year, dates were soon thrown into confusion. If, at a certain date, the summer solstice occurred on June 20th, in four years it would fall on the 21st; and thus it would gain one day every four years, until in time the summer solstice would happen in the winter months.

Julian Calendar.—Julius Cæsar first attempted to make the calendar year coincide with the motions of the sun. By the aid of Sosigenes, an Egyptian astronomer, he devised a plan of introducing every fourth year a leap-year, which should contain an extra day. This was termed a *bissextile year*, since the sixth (sextilis) day before the kalends (first day) of March was then counted twice.

Gregorian Calendar.—Though the Julian calendar was nearly perfect, it was yet somewhat defective. It considered the year to consist of 365¼ days, which is 11 minutes in excess. This accumulated year by year, until in 1582 the difference amounted to ten days. In that year, the vernal equinox occurred on the 11th of March, instead of the 21st. Pope Gregory undertook to reform the anomaly, by dropping ten days from the calendar and ordering that thereafter only centennial years which are divisible by 400 should be leap-years. The Gregorian calendar was generally adopted in Catholic countries. Protestant England did not accept the change until 1752. The difference had then amounted to 11 days. These were suppressed and the 3rd of September was styled the 14th.* Dates reckoned according to the Julian calendar are termed Old Style (O.S.); and those according to the Gregorian calendar, New Style (N.S.).

Commencement of the Year.—The Jews began their civil year with the autumnal equinox; but their ecclesiastical year, with the vernal equinox. When Cæsar revised the calendar, the Romans commenced the year with the winter solstice (Dec. 22), and it is probable he did not intend to change it materially. He ordered it to date from January 1, in order that the first year of his new calendar should begin with the day of the new moon immediately succeding the winter solstice.

The Earth our Timepiece.—The measure of time is, as we have just seen, the length of the mean day. This is estimated from the length of the sidereal day. Hence, the standard for time is the rotation of the earth on its axis. All weights and measures are based on time. An ounce is the weight of a given bulk of distilled water. This is measured

---

* This sweeping change was received in England with great dissatisfaction. Prof De Morgan narrates the following: " A worthy couple in a country town, scandalized by the change of the calendar, continued for many years to attempt the observance of Good Friday on the old day. To this end they walked seriously and in full dress to the church door, on which the gentleman rapped with his stick. On finding no admittance, they walked as seriously back again and read the service at home. There was a widespread superstition that, when Christmas day began, the cattle fell on their knees in their stables. It was asserted that, refusing to change, they continued their prostrations according to the Old Style. In England, the members of the Government were mobbed in the streets by the crowd, which demanded the eleven days of which they had been illegally deprived."

by cubic inches. The inch is a definite part of the length of a pendulum which vibrates seconds in the latitude of London. Arago remarks, a man would be considered a maniac who should speak of the influence of Jupiter's moons on the cotton trade. Yet there is a connection between these incongruous ideas. The navigator, travelling the waste of waters where there are no paths and no guide-boards, may reckon his longitude by the eclipses of Jupiter's moons, and so decide the fate of his voyage. We can easily see how the rotation of the earth on its axis influences the cost of a cup of tea.

VI. CELESTIAL MEASURE-MENTS.

Many persons read the enormous figures which indicate the distances and dimensions of the heavenly bodies with a questioning, indefinite idea, entirely unlike the feeling of certainty with which they read of the distance between two cities, or the number of square miles in a certain State. Many, too, imagine that celestial measurements are so mysterious in themselves that no common mind can hope to grasp the methods. Let us attempt the solution of a few of these problems.

1st. **To Find the Distances of the Planets from the Sun.**—In Fig. 108, E represents the earth; ES, the earth's distance from the sun; V, the planet Venus; and VES, the angle of elongation (a right-angled triangle). It is clear that, as Venus swings

apparently east and west of the sun, this angle may
be easily measured ; also, that it will be the greatest
when Venus is in aphelion and the earth in perihelion at the same time, for then VS will be the
longest and ES the shortest. Now in every right-

Fig. 103.

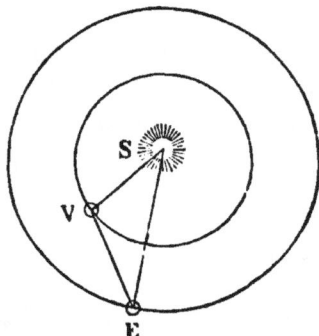

Comparative Distance of Venus and the Earth.

angled triangle the proportion between the hypothe-
nuse, ES, and the side opposite, VS, changes as the
angle at E varies, but with
the same angle remains the
same whatever may be the
length of the lines them-
selves. This proportion between the hypothenuse and
the side opposite any angle
is termed the *sine of that*

angle. Tables are published containing the sines for
all angles. In this way, the mean distance of Venus
is found to be $\frac{72}{100}$ that of the earth; Mars, $\frac{4}{3}$ times;
Jupiter, $5\frac{1}{4}$ times, etc.*

2nd. **To Measure the Moon's Distance from the
Earth.**—(1.) THE ANCIENT METHOD.—As the moon's
distance is so much less than that of the other
heavenly bodies, it is measured by the earth's semi-
diameter. The method, an extremely rough one,
which was in use among the ancients, was something

---

* If the pupil has studied Trigonometry, he may apply here the simple proportion—
ES : VS :: Radius : Sine of 47° 15″ = greatest elongation of Venus

The same result would be obtained by the use of Kepler's third law ; and on page
19, we saw how the distances of the planets themselves could be determined by the
periodic times, if the distance of the earth from the sun is first known. So that when
we have accurately determined the sun's distance from us, we can then decide by either
of the methods named the distance of all the planets. Indeed the sun's distance is, as
already remarked, the "foot-rule " for measuring all celestial distances.

like the following : In an eclipse of the moon, that
body passes through the earth's shadow in about
four hours. If, then, in four hours, the moon travels
along its orbit a distance equal to the diameter
of the earth, in twenty-four hours it would pass over
six times, and in a lunar month (about thirty days)
one hundred and eighty times, that distance. The
circumference of the lunar orbit, then, must be one
hundred and eighty times the diameter of the earth.
The ancients supposed the heavenly orbits to be
circles, and, as the diameter of a circle is about ⅓ of
the circumference, they deduced the diameter of the
moon's orbit as 120 times, and the distance of the
moon from the earth as 60 times, the semi-diameter
of the earth.

(2.) MODERN METHOD BY THE LUNAR PARALLAX.—
Under the head of parallax, we saw how, in common
life, we obtain a correct idea of the distance of an
object by means of our two eyes. We proved that
one eye alone gives no notion of distance. Just,
then, as we use two eyes to find how far from us an
object is, so the astronomer uses two astronomical
eyes, or observatories, located as far apart as possible, to find the parallax of a heavenly body. In
Fig. 109, M represents the moon ; G, an observatory
at Greenwich ; and C, another at the Cape of Good
Hope. At the former, the distance from the north
pole to the center of the moon, measured on a
meridian of the celestial sphere, is found to be 108°.
At the latter station, the distance from the south
pole to the moon's center is measured in the same
way, and found to be 73½°. The sum of these angles

is $181\frac{1}{2}°$. Now, the entire distance from the north pole around to the south pole, measured on a meridian, can be only half a great circle, or 180°. This difference of $1\frac{1}{2}°$ must be the difference in the position of the moon, as seen from the two observatories. For the observer at the former station will see the moon projected on the celestial sphere at G', and in measuring its distance from the north pole will measure an arc bG' further than if he were located at E, the center of the earth. The observer at the latter station will see the moon projected on the celestial sphere at C', and in measuring its distance from the south pole will measure an arc bC' more than if he were located at E, the center of the earth. The sum of bG' and bC' = G'C' is the difference in the position of the moon as seen from the two stations. In other words, it is the moon's parallax. The arc G'C', measures the angle C'MG'; that angle is equal to the opposite angle GMC = $1\frac{1}{2}°$. Now, in the four-sided figure GECM, the sides GE and CE are both equal radii of the earth = 3,956 miles; while the distance from G to C is the difference in the latitude of the two places. The angles ZGM and Z'CM, being the zenith distances of the moon, are known, and so the angles MGE and MCE are easily found. EM, the moon's distance from the center of the earth, is thus readily computed by a simple trigonometrical formula.

(3.) THE HORIZONTAL PARALLAX OF THE MOON is most commonly found by estimating its distance, not from the north and south poles, as just explained under the general meaning of the term parallax, but from a fixed star. The moon's horizontal parallax is now estimated at 57′, which makes its distance about sixty times the earth's semi-diameter (p. 273).*

Fig. 109.

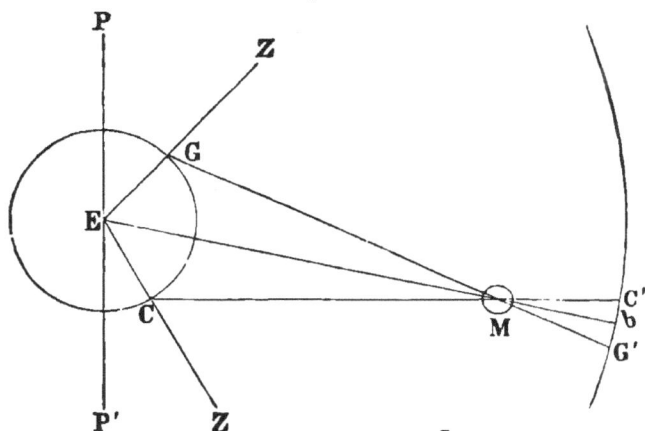

Measuring Moon's Distance from the Earth.

3. To Find the Sun's Distance from the Earth.— This might be estimated by obtaining the solar parallax in the same manner as the lunar parallax. It would be necessary only to take the sun's distance from the north and south poles respectively at Greenwich and the Cape of Good Hope, and then subtracting 180° from the sum of the two angular distances, the remainder would be the solar parallax. The difficulty in this method lies in the fact that when the sun shines the air is full of tremulous motion. This increases refraction—that plague of all astronomical calculations—to such an extent that it becomes im-

* In figure 110, let S represent the moon, sun, or any other heavenly body ; AB, the semi-diameter of the earth ; and ASB, the "horizontal parallax" of the body. Then, by the following trigonometrical formula, the distance from the earth may be easily calcu- lated—AS : AB :: Radius : Sin of ASB

possible to calculate so small an angle with any accuracy. Neither can the parallax be estimated, as in the case of the moon, by measuring the distance

Fig. 110.

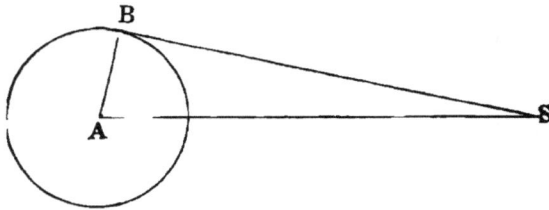

from a fixed star, since when the sun shines the stars near by are invisible even in a telescope. Astronomers have therefore been compelled to resort to other methods.

(1.) CALCULATION OF SOLAR PARALLAX BY OBSERVING MARS.—We have already seen that the distance of Mars from the sun is ⅔ that of the earth from the sun. If, therefore, we can find Mars's distance from the earth, we can multiply it by three, and so obtain the distance of the sun from the earth. In 1862, when Mars was in opposition, it came very near us, for it was in perihelion while the earth was in aphelion, so that its distance (as since ascertained) was only about 34,000,000 miles. Astronomers at Greenwich and the Cape, and at various American and European observatories, calculated the distance of the planet from the north and south poles, as well as from several fixed stars, in the manner just explained for obtaining the lunar parallax. The result of these observations fixed the solar parallax at 8".94,* making the sun's distance 91,430,000 miles.

* By the formula on page 275, we have, AS : AB :: Radius : Sin 8".94.

(2.) CALCULATION OF SOLAR PARALLAX BY OBSER-
VATION OF THE TRANSIT OF VENUS.—In the figure, let
A and B represent the position of two observers sta-

Fig. 111.

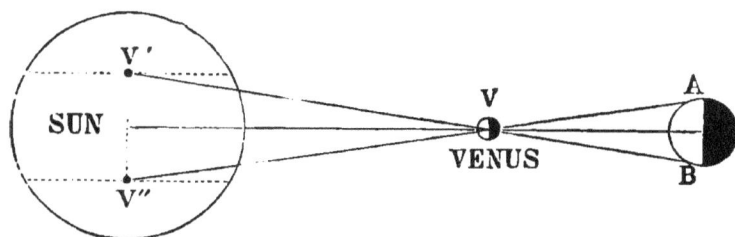

Transit of Venus.

tioned at opposite sides of the earth. At the time of
the transit, the one at A will see the planet Venus as
a round black spot at V" on the sun's disk, while the
one at B will see it at V'. The distance V'V" is the
difference in the position of Venus as seen from the
two stations on the earth. The distance AB is the
diameter of the earth. The distance V'V" is as much
greater than AB as VV" is greater than VA. The
distance of Venus from the sun is known, by Prob.
I., to be .72 that of the earth. The distance of Venus
from the earth must, then, be $1.00 - .72 = .28$. Hence,
VV", the distance from the sun to Venus, $= .72 \div .28 =$
2.5 times the length of AV, the distance of Venus from
the earth. Therefore, V'V" is equal to $2\frac{1}{2}$ times AB,
the earth's diameter, or 5 times the solar parallax. Knowing the hourly motion of Venus, it is
necessary only for each observer to find when the
planet's disk enters upon and leaves the sun's disk,
to determine the length of the path (chord) it
traces. A comparison of the length and direction

of these chords will give the length V′ V″ in seconds of space.

The advantage of this method is that, as the distance V′ V″ is two and half times that of AB, an error in measuring that chord affects the solar parallax less than one-fifth.

TIME OF A TRANSIT OF VENUS.*—This is an event of rare occurrence. It happens only at intervals of 8, 105½; 8, 121½, years, &c. Were the planet's orbit in the same plane as the ecliptic, a transit would take place during each synodic revolution; but as it is inclined about 3½°, the transit can occur only when the earth is at or near one of the nodes at the same time with the planet when in inferior conjunction. As the nodes of Venus now fall in that part of the earth's orbit which we pass in the beginning of June and December, transits always occur in those months.

THE TRANSIT OF JUNE 3RD, 1769, excited great interest. King George III. fitted out an expedition to Tahiti, under the command of the celebrated navigator, Capt. James Cook. In order to make the angle as great as possible, and so increase the length of the chords, or paths of the planet across the sun, astronomers were sent to all the most favorable points of observation—St. Petersburg, Pekin, Lapland, Cali

* The first transit ever seen was witnessed by Horrox, a young amateur astronomer residing near Liverpool. His calculations fixed upon Sunday, Nov. 24, 1639 (O. S.). He, however, commenced his watch of the sun on Saturday preceding. The following day he resumed his observation at sunrise. The hour for church arriving, he repaired to service as usual. Returning to his labor immediately afterward, he says: "At this time an opening in the clouds, which rendered the sun distinctly visible, seemed as if Divine Providence encouraged my aspirations; when—oh most gratifying spectacle! the object of so many earnest wishes—I perceived a new spot of perfectly round form that had just entered upon the left limb of the sun."

fornia, etc. They fixed the solar parallax at 8".58, making the sun's distance 95,293,055 miles.*

The transits of Dec. 8, 1874, and Dec. 6, 1882, were carefully observed by several government expeditions; the results have not yet been fully announced.

The next transits will happen,

June 8 . . . . 2004.
June 6 . . . . 2012.
December 11 . . . . 2117.
December 8 . . . . 2125.
June 11 . . . . 2247.

The transits of Mercury are more frequent; but owing to the nearness of the planet to the sun, they are of little value in determining the solar parallax.

CHANGES IN THE ESTIMATE OF THE SOLAR PARALLAX.—About 1824, Encke deduced 8".58 as the probable result of the observations upon the transit of 1769. This conclusion held the ground for nearly thirty years, and the corresponding solar distance of 95,293,000 miles is found in all the older text-books. About 1860, Le Verrier announced that he could reconcile the theories regarding certain of the planets only by assuming a greater solar parallax. As the result of various calculations, together with the material furnished by the observations upon the

* Le Gentil, sent out by the French Academy to observe the transit of 1761 in the East Indies, was prevented from making his first port by the war with England. High winds afterward kept him out at sea till the transit was over. He then resolved to remain abroad until after the transit of 1769. Eight long years passed, and the morning of June 3, 1769, dawned bright and beautiful. Le Gentil, with his instruments all in place, was counting the moments for the long-awaited transit to begin; when, suddenly, the sky grew black with clouds, and a tropical storm, the first in days, swept by. Meantime, Venus came and went, and the ill-fated Le Gentil had again lost the opportunity of years. Prostrated by his bitter disappointment, it was two weeks before he could hold his pen to write the story of his second failure.

planet Mars in 1862, a new parallax of 8".94 was obtained. This has been accepted by all until recently, and was used in former editions of this work. It is now known to be too large, and astronomers are making every effort to determine this most important factor in celestial measurements. As already stated on page 36, the parallax at present received is about 8".80, which represents a mean solar distance of 92,885,000; in round numbers, 93,000,000, as given in the present edition.

The difficulty of determining the solar parallax accurately will be seen, when one is told that the correction from the old value of 8".58 to the recent one of 8".94, was a change in the angle equal to that which the breadth of a human hair would make when seen at a distance of 125 feet. Yet this reduced the estimated distance of the sun from 95,293,000 miles, to 91,430,000 miles.

4. To Find the Longitude of a Place.*—(1.) THE SOLAR METHOD.—If the sailor can see the sun, he watches it closely with his sextant; and when the sun ceases to rise any higher in the heavens it is *apparent* noon. By adding or subtracting the equation of time (as given in his almanac), he obtains the true or *mean noon*. He then compares the local time thus determined, with the Greenwich time as kept by the

---

* It is pleasant to notice that the astronomer can *predict* with the utmost precision. He announces that on such a year, month, day, hour, and second, a celestial body will occupy a certain position in the heavens. At the time indicated, we point our telescope to the place, and, at the instant, true beyond the accuracy of any timepiece, the orb sweeps into view! A prediction of the Nautical Almanac is received with as much confidence as if it were a fact contained in a book of history. "On the trackless ocean, this book is the mariner's trusted friend and counsellor; daily and nightly its revelations bring safety to ships in all parts of the world. It is something more than a mere book. It is an ever-present manifestation of the order and harmony of the universe."

ship's chronometer. The difference in time reduced to degrees, gives the longitude.

(2.) THE LUNAR METHOD.—On account of the difficulty in obtaining a watch which will keep the exact Greenwich time through a long voyage, the moon is more generally relied upon than the chronometer. The Nautical Almanac is always published, for the benefit of sailors, three years in advance. It gives the distance of the moon from the principal fixed stars which lie along its path, at every hour in the night. The sailor has only to determine with his sextant the moon's distance from any fixed star, and then, by referring to his almanac, find the corresponding Greenwich time. By comparing this with the local time, and reducing the difference to degrees, etc., he obtains the longitude.

5. To Find the Latitude of a Place.—(1.) By means of the sextant find the elevation of the pole above the horizon, and this gives the latitude directly. (Fig. 35.)

(2.) In the same manner, determine the height of the sun above the horizon at noon. The sun's declination for that day (as laid down in the almanac), added to or subtracted from this, gives the height of the equinoctial above the horizon. Subtract this result from 90°, and the remainder is the latitude.

"Place an Astronomer on board a ship; blindfold him; carry him by any route to any ocean on the globe, whether under the tropics or in one of the frigid zones; land him on the wildest rock that can be found; remove his bandage, and give him a chronometer regulated to Greenwich or Washington time, a transit instrument with the proper appliances, and the necessary books and tables, and in a single clear night he can tell his position within a hundred yards by observations of the stars."

6. To Find the Circumference of the Earth.—If the earth were a perfect sphere, it is obvious that degrees of latitude would be of the same length wherever measured on its surface. Each would be $\frac{1}{360}$ of the entire circumference. If, however, a person sets out from the equator, and travels along a meridian toward either pole, and, when the polar star has risen in the heavens one degree above the horizon, he marks the spot, and then continues his journey, marking each degree in succession, he will find that the degrees are not of equal length, but increase gradually from the equator to the pole. If, now, the length of a degree be measured at different places, the rate of variation can be found, and then the *average* length be estimated. Measurements for this purpose have been made in Peru (almost exactly at the earth's equator), Lapland, England, France, India, Russia, etc. So great accuracy has been attained, that Airy and Bessel, who have solved the problem independently, differ in their estimate of the equatorial diameter but 77 yards, or only $\frac{44}{1000}$ of a mile.

7. To Find the Relative Size of the Planets.—The volumes of two globes are proportional to the cubes of their like dimensions. The diameter of Mercury is 3,000 miles, and that of the earth 7,925; then,

The volume of Mercury : the volume of the earth : : 3000^3 : 7925^3.

The same principle applied to the volume or bulk of the sun gives—

The bulk of the sun : bulk of the earth :: $866,000^3$: 7925^3.

8. To Find the Diameter of the Sun.—(1.) A very simple method is to hold up a circular piece of paper before the eye at such a distance as exactly to hide the entire disk of the sun. Then we have the proportion,

As dist. of paper disk : dist. of sun's disk : : diam. of paper d. : diam. sun's d.

(2.) The apparent diameter of the sun, as seen from the earth, is about 32': the apparent diameter of the earth, as seen from the sun, is twice the solar parallax, or 17".60 (p. 36). Thence, the

Ap. diam. of earth : ap. diam. of sun : : real diam. of earth : real diam. of sun.

(3.) Knowing the apparent diameter of the sun, and its distance from the earth, the real diameter is found by Trigonometry. In figure 110, let S represent the earth ; AB, the radius of the sun ; and ASB, half the apparent diameter of the sun. We shall then have the proportion,

AS : AB : : radius : sin. 16' (half mean diam. of sun).

By a similar method the diameters of the planets are obtained.

PRACTICAL QUESTIONS.

1. In what constellation is Job's Coffin ? The Letter Y ? The Scalene Triangle ? The Dipper ? The Kids ? The Triangles ?
2. Name some facts in the solar system for which the nebular hypothesis fails to account.
3. Which is probably hotter, a yellow or a red star ?
4. Are any of the stars likely to collide with each other ?
5. Is the real day longer or shorter than the apparent one ?
6. Do we ever see the stars ?

7. What fixed star is nearest the earth ?

8. How often is Polaris on the meridian of a place?

9. How do we know that the stars are suns ?

10. Can a watch keep apparent time ?

11. How could a child be 8 years old before a return of its birthday ?

12 When will a watch and a sun-dial agree ?

13. What star will be the Pole Star next after Polaris?

14. Why is the birthday of Washington celebrated on Feb. 22, when he was born Feb. 11, 1732 (O. S.) ?

15. Does the tide have any effect on the length of the day ?

16. Will the Big Dipper always look as it does now ?

17. How many times does the earth turn on its axis every year ?

18. Does the spectroscope tell us anything concerning the constitution of the moon, or any of the planets ?

19. When the United States bought Alaska from Russia, the calendar used there was found to be one day ahead of our reckoning. Why was this ?

20. Why do the dates of the solstices and equinoxes vary a day in different years ?

21. Why are not forenoon and afternoon of the same day, as given in the almanac, of equal length ?

22. In what part of the heavens do the stars apparently move from west to east ?

23. What year was only nine months and six days long ?

24. What day will be the last day of the Nineteenth Century ?

25. If one should watch the sky, on a winter's evening, from 6 P. M. to 6 A. M., what portion of the celestial sphere would he be able to see ?

26. How do we know that the moon has little, if any, atmosphere ?

27. In Greenland, at what part of the year will the midnight sun be seen due north ?

28. Can you give any other proof of the rotundity of the earth, besides that named in the text ?

29. Point out the error in the following passage from Byron's "Darkness" where the poet, in describing the effect of the sun's destruction, says —

"I had a dream,   *   *   *
        *   *   *   which was not all a dream,
The bright sun was extinguished, and the stars
Did wander darkling in the external space
Rayless and pathless."

30. Explain the remark of the First Carrier in Scene I, Act II, King Henry IV: "An't be not four by the day, I'll be hanged : Charles' wain is over the new chimney."

31. Why does not the earth move with equal velocity in all parts of its orbit?

32. How many Jovian-years old are you?

33. Why is the sky blue?

34. At what season of the year does Christmas occur in Australia?

35. What causes the apparent movement of the sun north and south?

36. On what part of the earth is the twilight the longest? The shortest?

37. Name the causes which make our summer longer than winter.

38. Why is not total darkness produced when a dense cloud passes between us and the sun?

39. Why does the time of the tide vary each day?

40. Why is an annular, longer than a total, eclipse?

41. Why is it colder in winter than in summer?

42. Do the solar spots affect our weather?

43. Can the moon be eclipsed in the day-time?

44. Why are the sidereal days of uniform length?

45. Why are not the solar days of uniform length?

46. What does the moon's phases prove?

47. Why do the sun and moon appear flattened when near the horizon?

48. How many stars can we see with the naked eye?

49. Is there ever an annular eclipse of the moon?

50. "While the sun rises and sets 365 times, a star rises and sets 366 times." Explain.

51. How many moons are there in the solar system?

52. What causes the twinkling of the stars?

53. Name some of the uses of the stars. *

* "To the astronomer, the fixed stars are immovable boundary-stones by which he determines the courses of the wandering heavenly bodies. To the geographer, they are the signal-stations according to which he surveys the chart of the earth by the heavens. To the mariner, they are the lights that direct him over the dark paths of the seas. To the hunter, the herdsman, the wanderer, they are a clock. To the farmer, they are a calendar. The historian finds in them many a memorable event in the oldest Grecian history. The poet reads in them the charming Grecian mythology, which has furnished such rich materials to dramatic art ; and every person of sensibility receives from them an impulse to worship, meditation, and hope."

54. Describe the methods by which we determine the distance of the sun from the earth.

55. Why do not the signs and the constellations of the Zodiac agree?

56. When we look at the North Star, how long since the light that enters our eye has left that body?

57. In what direction does a comet's tail generally point?

58. What is the cause of shooting stars?

59. Why does the crescent moon appear larger than the dark body of the moon?

60. What is the real path of the moon?

61. What would be the result if the axis of the earth were parallel to the plane of its orbit?

62. Do we see the same stars at different seasons of the year?

63. Why do we not perceive the earth's motion in space?

64. Did the earth ever shine as a star? Does it now shine as a planet?

65. What is the nebular hypothesis?

66. What is the cause of the solar spots?

67. Would it make the new moon "drier" or "wetter" if the moon's path ran north of, instead of on, the ecliptic at the time of new moon?

68. Under what conditions are we accustomed to transfer motion?

69. Why do not the planets twinkle?

70. Why is the horizon a circle?

71. What causes are gradually increasing the length of the day?

72. What distance does the moon gain in her orbit each year?

73. State the general argument which renders it probable that other worlds are inhabited.

74. Illustrate the uniformity of Nature. What thought does this suggest?

75. At what rate are we traveling through space? How is this determined?

76. Why does the length of a degree of latitude increase in going from the equator toward either pole of the earth?

77. How can you detect the yearly motion of the sun among the stars?

78. Have you actually traced the movement of any one of the planets, so as to understand its peculiar and irregular wandering among the stars?

79. How do you explain the varied aspect of the heavens in the different seasons of the year?

80. How does the spinning of a top illustrate the subject of precession ?

81. Why do solar eclipses come on from the west and cross to the east, while lunar eclipses come on from the east and cross to the west ?

82. Newcomb, in his Astronomy, says that, "If, when the moon is near the meridian, an observer could in a moment jump from New York to Liverpool, keeping his eye fixed upon that body he could see her apparently jump in the opposite direction about the same distance." Explain.

83. When, and by whom, was the basis of the calendar we now use fully established ?

84. How much is the Russian reckoning of time behind ours ?

85. Is there any gain in having the astronomical and the calendar year agree ?

86. What religious festival is fixed each year by the motion of the moon ?

87. Why can we, at different times, see *both poles* of the planet Mars ?

88. What famous astronomical discovery was made on the first day of this century ?

89. Do the stars rise and set at the poles ?

90. Name and locate the stars of the first magnitude which are seen in our sky.

91. Name three bright stars which lie near the first meridian.

92. What events were transpiring in our history a Saturnian century ago?

93. What is the sun's declination at the winter solstice ? At the autumnal equinox ?

94. Will the width of the terrestrial zones always remain exactly as now ?

95. Is it always noon at 12 o'clock ?

96. When the sun's declination is $23\frac{1}{2}°$ N., in what sign is he then located, and what is his R. A. ?

97. What is the apparent diameter of the sun ?

98. How can a sailor find his latitude and longitude at sea ?

99. How many miles on the solar disk represent a second of apparent diameter ?

100. At what latitude will there be twilight during the entire midsummer night ?

Fig. 113.

Cambridge Equatorial Telescope.

IV.

APPENDIX.

APPENDIX.

TABLE ILLUSTRATING KEPLER'S THIRD LAW. (CHAMBERS.)

IN the first column are the relative distances of the planets from the sun; in the second, the periodic times of the planets; and in the third, the squares of the periodic times divided by the cubes of the mean distances. The decimal points are omitted in the third column for convenience of comparison. The want of *exact uniformity* is doubtless due to errors in the observations.

Vulcan ? - - - - - - - - -	.143	19.7	132 716
Mercury - - - - - - - -	.38710	87.969	133 421
Venus - - - - - - - - -	.72333	224.701	133 413
Earth - - - - - - - - -	1.	365.256	133 409
Mars - - - - - - - - -	1.52369	686.979	133 410
Jupiter - - - - - - - -	5.20277	4,332.585	133 294
Saturn - - - - - - - -	9.53858	10,759.220	133 375
Uranus - - - - - - - -	19.18239	30,686.821	133 422
Neptune - - - - - - - -	30.03627	60,126.722	133 413

Arago, speaking of Kepler's Laws, says: "These interesting laws, tested for every planet, have been found so perfectly exact, that we do not hesitate to infer the distances of the planets from the sun from the duration of their sidereal periods; and it is obvious that this method possesses considerable advantages in point of exactness."

MEASUREMENTS OF THE EARTH'S DIAMETER.

	AIRY.	BESSEL.
Polar diameter - - - - - - - - - - -	7899.17	7899.11
Equatorial diameter - - - - - - - - -	7925.64	7925.60
Compression - - - - - - - - - - -	26.47	26.49

TABLE OF THE PERIODIC COMETS.

Compiled by Lewis Swift, F.R.A.S., Director of Warner Observatory, Rochester, N. Y.

Name of Comet.	Periodic time. (Years.)	Perihelion distance.	Aphelion distance.	Dist. from node to perihelion.	Longitude of node.	Inclination.	Eccentricity.	Direction of motion.	Next return.
Encke's	3.303	0.342	4.10	183 56	334 34	12 53	0.8455	D	1885
Tempel's (I)	5.20	1.34	4.66	185 7	121 1	12 46	0.5536	D	1889
Swift's	5.50	1.07	5.14	106 13	296 51	5 23	0.6553	D	1886
Brorsen's	5.56	0.62	5.68	14 55	101 20	29 23	0.8098	D	1884
Winnecke's	5.64	0.78	5.50	165 12	111 31	11 17	0.7406	D	1886
Tempel's (II)	6.00	1.77	4.82	159 25	78 46	9 46	0.4630	D	1885
D'Arrest's	6.39	1.17	5.72	173 0	146 19	15 43	0.6278	D	1890
Biela's (North)	6.59	0.96	6.17	109 20	246 05	12 33	0.7552	D	—
Biela's (South)	6.63	0.86	6.20	100 13	216 09	12 34	0.7561	D	—
Faye's	7.41	0.69	5.92	200 15	209 42	11 22	0.5574	D	1888
Denning's	9.00 ?			312 39	65 54	6 50	0.8940	D	1891 ?
Tuttle's	13.78	1.03	10.51	206 47	269 17	54 17	0.8210	D	1885
Pons-Brook's	71.34	0.77	33.41	199 11	254 5	74 2	0.9549	D	1955
Halley's	76.37	0.59	35.30	112 43	57 15	162 15	0.9675	R	1912

QUESTIONS FOR CLASS USE.

THESE are the questions which the author has used in his own classes for review and examination. In the historical portion, he has required his pupils to write articles upon the character and life of the various persons named, gathering materials from every attainable source. He has also introduced whatever problems the class could master, taking topics from the article on Celestial Measurements and the various mathematical treatises.

INTRODUCTION.—Define Astronomy. Is the earth a planet? Is the moon a planet? What is the sky? Why does it seem concave? What gives it its color? What is the difference in the appearance of a fixed star and a planet? What is the Milky Way? In what direction does it span the heavens? In what season of the year is it most brilliant?

I. THE HISTORY.

5–6. What can you say of the antiquity of astronomy? How far back do the Chinese records extend? Name some astronomical phenomena they contain. Why were these astronomers at fault in failing to announce the eclipse? (Ans. Certain religious ceremonies were performed on such occasions, and their omission was believed to expose the nation to the anger of the gods.) Why should the Chaldeans have become versed in this study? How ancient are their records? What discoveries did they make? How does the Asiatic differ from the European mind?

7. What Grecian philosopher early acquired a reputation in this science? What other discovery did Thales make (Physics, p. 251)? What did he teach? What memorable eclipse did he predict? What did Anaximander teach? In what century did Pythagoras live? What was his characteristic trait? Did he advance any proof of his system? Explain his theory. How does it differ from ours? What strange views did he hold?

8. When did Anaxagoras live? What did he teach? What theory did Eudoxus advance? What is the theory of the crystalline spheres?

What has Hipparchus been styled? What addition did he make to astronomical knowledge? How many stars in our present catalogue (p. 207)? How did Egypt rank in science at an early day? What preparation did the Grecian philosophers make to fit themselves for teachers? How long did Pythagoras travel for this purpose?

9. What can you say of the School at Alexandria? What great work did Ptolemy write? What theory did he expound? Was it original? What discovery did Eratosthenes make? Describe that method (p. 282). Show how the movements of the planets puzzled the ancients.

10. What was the theory of "cycles and epicycles"? Did the ancients believe in the reality of this cumbrous machinery? Did this theory possess any accuracy?

11. Could it be adapted to explain any new motion? What was the remark of Alfonso? Describe the progress of learning among the Saracens.

12. Where was the first Observatory in Europe built? When did Spain lose her prominence in scientific studies?

13. What is astrology? What was its association with astronomy? State something of the repute in which astrology was held. Tell what you can of the system. What use did it subserve?

14. What theory displaced the Ptolemaic? When? Was the system of Copernicus original? What credit is due him? Describe his idea of apparent motion. How did he apply this to the heavenly bodies? What crudity did he retain?

15. Who was Tycho Brahe? What was his theory? How did it differ from Ptolemy's and Copernicus's? What good did Brahe accomplish? Could he generalize his facts? Had he a telescope? How did Kepler differ from Brahe? What were the two prominent characteristics of Kepler?

16-19. State his three laws. Tell how he discovered the first. The second. The third. Describe the ellipse. Define focus, perihelion, and aphelion. What remarkable statement did Kepler make? When did Galileo live?

20. What discoveries did he make in Physics? In astronomy? What advantage did he have over his predecessors? Give an account of his observations on the moon. On Jupiter's moons.

21. Why did this settle the controversy between the Ptolemaic and the Copernican system? How were Galileo's discoveries received? Give some of Sizzi's arguments. Who discovered the law of gravitation?

23. Repeat it. How was this idea suggested? What familiar laws of motion aided Newton? How did he apply these to the motion of the moon? Repeat the story of his patient triumph.

24, 25. What is the celestial sphere? Give the two illustrations which show its vast distance from the earth. Why can we not see the stars by day, as by night? What portion of the sphere is visible to us? Name the three systems of circles.

26–30. Name and define (1) the principal circle, (2) the secondary circles, (3) the points, and (4) the measurements of each system. Define especially, because in common use, zenith, nadir, azimuth, altitude, equinoctial, right ascension, declination, equinox, ecliptic, colure, and solstice. What is N or S in the heavens?

31. What is the Zodiac? How wide is it? How ancient? How is it divided? State the names and signs.* State the meaning of each (p. 210.)

II. THE SOLAR SYSTEM.

What bodies compose the solar system? Describe how we are to picture it to ourselves.

THE SUN.—What is its sign? Its distance from us? Illustrate. What is the solar parallax (see pp. 121, 275)? What change has recently been made in the estimate of the parallax of the sun, and of its distance from the earth? (See p. 279.)

37. How are celestial distances measured? What is the color of the sun? To what is the sun's light equal? To how many moons?

38. To what is its heat equal? Illustrate. What proportion of the sun's heat reaches the earth? What is the apparent size of the sun? How does this vary?

39, 40. State the solar dimensions. (1) diameter—illustrate; (2) volume; (3) mass; (4) weight; (5) density. How large did Pythagoras think the sun is? Tell something about the force of gravity on the sun. How much would you weigh if carried to its surface? (The force of gravity on the sun as compared with the earth is 27.6.) How does the sun appear to the naked eye?

41. How can we see the spots? What were formerly the views of astronomers with regard to the sun's face?

42. When were the spots discovered? Tell something about the number of the spots. Their location. Size. What number of miles subtend a second of arc at the distance of the sun?

* "The Ram, the Bull, the Heavenly Twins,
And next the Crab, the Lion shines,
The Virgin and the Scales,
The Scorpion, Archer, and He-goat,
The Man that bears the watering-pot,
And Fish with glittering tails."

43. Describe the parts of which the spots are composed. Describe the motion of the spots.

44. How do the spots change in form as they pass across the disk? What does this prove? What is the length of a solar axial rotation?

45, 46. Explain a sidereal and a synodic revolution of a spot. Why do not the spots move in straight lines? Show how they curve. Tell what you can about the irregular movements of the spots.

47. Illustrate how suddenly they change. What can you say about their periodicity? Who discovered this? Is there any connection between the solar spots and the aurora?

48. Tell about the influence of the planets on the spots. Do the spots affect the fruitfulness of the season? Does the temperature of the spots differ from that of the rest of the sun? Are the spots depressions in the sun?

49. How much darker are they than the adjacent surface? Is the sun brighter than the Drummond light? (*Ans.* "The sun gives out as much light as one hundred and forty-six lime-lights would do, if each were as large as the sun and were burning all over.")

50. What are the faculæ? Describe the mottled appearance of the sun. What is the shape of the bright masses? What is a granule? What is its size?

51, 52. Describe the constitution of the sun according to Wilson's theory. How are the spots produced? The faculæ? The penumbra? The nucleus? The umbra?

53, 54. What is the present theory ("Kirchhoff's Theory")? Name the four different portions of the sun. Define the nucleus. The photosphere. The chromosphere. The corona. What are the protuberances? How are the spots produced? The umbra? The penumbra? Upon what discoveries does this theory depend (p. 262)? What is the cause of the heat of the sun? Will the heat ever cease?*

THE PLANETS.—Name the six characteristics common to all the planets. Compare the two groups of the major planets.

57, 58. Draw an ellipse, and name the various parts. Define the ecliptic. The plane of the ecliptic. Why is the ecliptic so called? Define the ascending node. The descending node. Line of the nodes. Longitude of the node. Tell what you can with regard to the comparative size of the planets.

* If we accept the Nebular hypothesis (p. 256), we must suppose that the heat is produced by the condensation of the nebulous matter and consequent chemical changes. The sun is radiating its heat constantly, and, at some time, its light will go out, in turn, as that of the earth and the planets has before it. This theory is of especial interest, as it shows that the sun, as well as the solar system, has a certain fixed existence; and that, "like all natural objects, it passes through its regular stages of birth, vigor, decay, and death, in one order of progress."—*Newcomb.*

60. What is a conjunction? Name the earliest that are recorded.

61-3. What do you say concerning the probability of the planets being inhabited?* State the conditions of life on the different planets. What are the two divisions of the planets?

64. What causes the apparently irregular movements of the planets? Define heliocentric and geocentric places. Illustrate. In what part of the sky is an inferior planet always seen? Define inferior and superior conjunction. Greatest elongation.

65. Why is a star at one time " evening " and, at another, " morning star"? What is a transit?

65, 66. Explain the retrograde motion of an inferior planet. (This motion, it will be remembered, was one that sorely puzzled the ancients.) Describe the phases of an inferior planet.

67. Why does an inferior planet have phases? Define gibbous. Explain the opposition and conjunction of a superior planet.

68. Explain its retrograde motion. Must a superior planet always be seen in the same part of the sky as the sun? Define quadrature. Can an inferior planet be in quadrature?

69, 70. Which retrogrades more, a near or a distant planet? Define a sidereal and a synodic revolution of an inferior and a superior planet, and tell what you can about each. In what case would there be no difference between a sidereal and a synodic revolution? Why is a planet invisible when in conjunction? When is a planet evening, and when morning star?

71. Tell what you can about the supposed discovery of a planet interior to Mercury. What name and sign have been given to it?

MERCURY.—Definition and sign? Describe the appearance of Mercury, and where seen. What was the opinion of the ancients? Of the astrologists? Of chemists? Why is it difficult to see this planet? When can we see it best?

73. What is the peculiarity of its orbit? What is Mercury's greatest elongation from the sun? Why does this vary? What is Mercury's distance from the sun? What is its velocity? What is the length of

---

* The uniformity of Nature is a most effective argument in this direction. Light travels everywhere through the universe at the same rate. The elements of star, planet, and the earth are the same. The sun, which may be considered as the mother of the earth, is composed of the same materials. The laws of gravitation rule so absolutely that the satellite of Sirius was not discovered until after it was observed that an unknown influence affected the star. "The uniformity of law and matter is proof that there must be through the universe organizations similar to those of our system. We see the result of these laws in the world we inhabit, and we cannot doubt that the same powers and the same materials have produced organizations similar to these of the earth in millions of other places, though we can only philosophically suppose their existence, not practically prove it."—W. Meyer.

its day? Of its year? What is the difference between its sidereal and synodic revolutions? What is its distance from the earth?

74-6. Show why its greatest and least distances vary so much. What is its diameter? Volume? Density? Force of gravity? Specific gravity? How much would you weigh on Mercury? (Mercury's force of gravity as compared with that of the earth is .46.) Describe its seasons. (If the pupil does not understand pretty well the subject of the terrestrial seasons, it would be well here to read carefully page 95, et seq.) What is the temperature? The appearance of the sun? Has Mercury any moon? What is the appearance of the planet through a telescope? What do these phases prove? What do we know of the mountains and valleys upon Mercury? The atmosphere? Have we any recent observations?

77. VENUS.*—Definition and sign? Ancient names? Appearance to us? When brightest? Can Venus be seen by day? Illustrate.

78. Describe the orbit of Venus. What is the distance of Venus from the sun? Velocity? Length of the year? Day? Difference between the sidereal and synodic revolutions? Distance from the earth?

79. How near may Venus approach us? How does her apparent size vary? When is Venus the brightest? What is her diameter? Volume? Density?

80. Force of gravity? (The force of gravity on the surface of Venus is .82 that of the earth.) Does the force of gravity increase or decrease with the mass or volume of the body? Describe the seasons upon Venus.

81, 82. Describe the telescopic appearance of Venus. Who discovered the phases of Venus? What was the effect of this discovery? What proof have we that Venus possesses a dense atmosphere? Has Venus a moon?

83. EARTH.—Sign? What is the appearance of the earth from the other planets? Do we, then, live on a star? Is it probable that the earth was always dark and dull as it now seems to us?* How does

---

* Venus is the only planet mentioned by Homer—

Οἷος δ'ἀστὴρ εἶσι μετ' ἄστρασι νυκτὸς ἀμολγῷ
Ἕσπερος ὃς κάλλιστος ἐν οὐρανῷ ἵσταται ἀστήρ
                                    Iliad, xxii. 317.

† Probably not. The earth was doubtless once a glowing star, like the sun. Its crust is only the ashes and cinders of that fearful conflagration. The rocks are all burnt bodies. The atmosphere is only the gas left over after the fuel was all consumed Every organic object has been rescued by plants and the sunbeam from the grasp of oxygen.

the size of the earth compare with that of the other planets? What is the shape of the earth? What is its exact diameter? (See Table in the Appendix.)

84. Circumference? Density? Weight? What comparison may be made to illustrate its inequalities? How do you prove the rotundity of the earth? *

85. Why can we see further from the top of a hill than from its base? Why is the horizon a circle?

86-7. Give some illustrations of apparent motion. Is it, then, natural for us to *transfer* motion? Under what conditions do you think this occurs? Explain the cause of the rising and setting of the sun and stars. Who first explained these phenomena in this manner? What do you say of its simplicity?

88. What is the cause of day and night? Do all places on the earth revolve with equal velocity? Illustrate. At what rate do we move? Why do we not perceive our motion?

89. What would be the effect if the earth were to stop its rotation? Is there any danger of this catastrophe? How is the length of the day increasing? Is the amount appreciable?

90-1. Draw the figure, and show how the stars move daily through unequal orbits and with unequal velocities. Describe the appearance of the stars at the N. Pole. At the Equator. At the S. Pole.

92-3. Describe the path of the earth about the sun. Defin eccentricity. What is the amount of the eccentricity of the earth's orbit? Is this stable? Do we see the same stars at different seasons of the year? Why not? If we should watch from 6 P. M. to 6 A. M., what portion of the sphere would we see?

94. What do we mean by the yearly motion of the sun among the stars? How can we see it? What is the cause? What is the ecliptic? Why so called? What are the equinoxes? What do you understand when you see in the almanac the statement that " The earth is in Aries?" "The sun is in Sagittarius?" etc. How many apparent motions has the sun? Name them, and give the cause and effects of each. Has the sun any real motions?

95. Describe the apparent motion of th sun, N. and S. How is it that the sun in summer shines on the north side of some houses both at rising and setting, but in winter never does? Define the obliquity of the ecliptic. The parallelism of the earth's axis. What do you say of its permanence? Why will a top stand while spinning, but will fall as soon as it ceases?

97. Show how the rays of the sun strike the various parts of the

* It is said that aeronauts, at a proper height, can distinctly see the curving form of the earth's surface.

earth at different angles at the same time. Show how the angles vary
at different times. Is the sun really hotter in summer than in winter?
Why does it seem to be? Why is it warmer in summer than in winter? What effect upon the temperature has the difference in the length
of the summer and the winter day?

98–100. Explain the cause of equal day and night at the equinoxes.
Why are our days and nights of unequal length at all other times?
Why does the length vary at different seasons of the year? How do
the seasons, &c., in the N. Temperate Zone compare with those in the
S. Temperate Zone? Describe the yearly path of the earth about the
sun—(1), at the summer solstice ; (2), at the autumnal equinox ; (3), at
the winter solstice ; (4), at the vernal equinox ; (5), the yearly path
finished back to the starting-point. Is the division of the earth's
surface into zones an artificial or a natural distinction? Who invented it?

101. How much nearer are we to the sun in winter than in summer?
Why is it not warmer in winter? How is it in the South Temperate
Zone? When do the extremes of heat and cold occur? Why do they
not occur exactly at the solstices?

102. Why is summer longer than winter? Does the earth move
with the same velocity in all parts of its orbit? Describe the curious
appearance of the sun at the North Pole. In Greenland, at what part
of the year will the midnight sun be seen due north? What is the
length of the days and nights at the Equator?

103. Describe the results if the axis of the earth were perpendicular
to the ecliptic. If the equator were perpendicular to the ecliptic.

104–5. Define the precession of the equinoxes. Who discovered
this fact? At what rate does this movement proceed? What time will
be required for the equinoxes to make an entire revolution? What
are the results of precession? What star was formerly the Polar Star?
(See p. 219.)

106–9. Explain the cause of precession. How does the spinning of a
top illustrate this subject? In what way is the force which acts on a
spinning-top opposite to that which produces precession? What is
Nutation? What is the cause of the nodding motion? How does the
moon's influence compare with that of the sun? What is the effect of
Nutation?

110. What is the real path of the N. Pole through the heavens? Is
the obliquity of the ecliptic invariable? What is the period of this
oscillation?

111. What causes combine to produce this nodding motion we have
described? Why are the tropics located where they are? Is their
position on the earth permanent? What effect does precession have

on the latitude of the stars? What is meant by the line of apsides of the earth's orbit? Is this permanent?* What is the Great Year of the Astronomers?

112. At what season of the year is the earth now in perihelion? When was it in perihelion in the autumn? When in the winter? When will perihelion occur in the spring? When in summer? When will the cycle be completed? What provision is there for permanence in the midst of these changes?

113-14. What is refraction? Its effect? Upon what principle of Optics is this based? How does refraction vary? Are the sun and moon ever where they seem to be? Is the real day longer or shorter than the apparent one?

115. Describe the apparent deformation of the sun and moon near the horizon. Explain. Why are not these bodies apparently deformed in the same way when they are high in the heavens? Why do they appear smaller in the latter case? (See Fig. 48, p. 124.) What causes the hazy appearance of the heavenly bodies near the horizon?

116. What is the cause of twilight? How long does it last? Is it the same at all seasons of the year? In all parts of the earth?

117. Where is it the longest? Shortest? State the cause of this variation. What is diffused light? What would be the effect if the atmosphere did not act in this way? Is there really any sky in the heavens? What is the cause of the appearance? What is aberration of light?

118. Illustrate this phenomenon. State two reasons why we never see the sun where it really is.

119. What is the general effect of aberration? Define parallax. Illustrate.

120-21. Define true and apparent place. How does parallax vary? What is the practical importance of this subject (pp. 36, 278)? Define horizontal parallax. What is the sun's horizontal parallax? What is the annual parallax?

THE MOON.—Signs? Describe the moon's orbit. What is the moon's distance from the earth? Illustrate. What is the difference between her sidereal and synodic revolutions? What is the real path of the moon? (Imagine a pencil fastened to the spoke of a wheel, and the wheel rolled by the side of a wall on which the pencil is constantly marking.) How often does the moon turn on her axis? What is the

* "The line of equinoxes of the earth's orbit. as we have seen, has a slow *left-handed retrograde motion* of 50".2 each year, called the precession of the equinoxes; and the line of apsides has a still slower *right-handed direct motion* of 11".29; and in consequence of the motion of both these lines, the angle formed by them changes through 61".49 each year, so as to complete an entire revolution in 21,077 years."

moon's diameter? Volume? How does her apparent size vary? Why does she appear larger than she really is?

124. Why does the crescent moon appear larger than the dark body of the moon? When ought the moon to appear the largest? Do all persons think the moon to be of the same apparent size?

125. Explain the three librations of the moon. How does moonlight compare with sunlight? Is there any heat in moonlight?

126. Does the center of gravity in the moon coincide with that of magnitude? Has the moon any atmosphere? What proof have we of this? (*Ans.* (1). We see but slight, if any, appearance of twilight on the moon. (2). When the moon passes between us and a star, it does not refract the light of a star, so that the atmosphere cannot be sufficient to support more than $\frac{3}{100}$ of an inch of the mercurial column.) What must be the effect of this lack upon the temperature of the moon's surface? State Langley's observations upon Mount Whitney. How does the earth appear from the moon?

127-9. What is the earth-shine? How is it caused? What is it called in England? Describe the path of the moon around the earth, and the consequent phases. Why is new moon seen in the west and full moon in the east? Why can we sometimes see the moon in the west after the sun rises, and in the east before the sun sets? What is the length of a lunar month?* What do we mean by the moon's running high or low? What is the cause of this variation? Is it of any use?

130-1. What is harvest moon? What is the cause?†

132. Explain the cause of "Dry Moon" and "Wet Moon." What are nodes? How much is the moon's orbit inclined to the ecliptic— our ideal sea-level? What is an occultation? What use does it subserve? Describe the seasons, heat, etc., on the moon.

135-7. Describe the telescopic appearance of the moon. Are the mountains the light or the dark portions? What can you say about them? The gray plains? The rills? The craters? What are the peculiar features of the lunar landscapes? Are the lunar volcanoes extinct?

---

* "The moon's sidereal period is not constant, and a comparison of modern with ancient observations shows that it has undergone an acceleration since the period of the Chaldean observations of eclipses made 720 B. C. Several explanations have been given by Laplace and others, of the supposed cause of the acceleration of the moon's mean motion; but it is highly probable that it is a *pseudo-phenomenon*, that owes its origin to a real lengthening of the time of rotation of the earth (which is the unit of astronomical time), caused by the friction of the sea and atmosphere."

† It will aid in understanding the cause of harvest moon, if one gets clearly in mind the fact that the moon when full is always in the opposite part of the heavens from the sun. At the time of the autumnal equinox, *i. e.* when the sun is at the autumnal equinox, (or in Libra, note, p. 94,) the moon must be at the vernal equinox, (or in Aries.) The least retardation of the moon, which occurs at this time, happens, therefore, in September.

138. ECLIPSES.—When can an eclipse of the sun occur? Show how a solar eclipse may be total, partial, or annular. Define umbra. Penumbra. Central eclipse. State the general principles of a solar eclipse. What curious phenomena attend a total eclipse?* What are Baily's Beads? What is the corona? Describe the effect of a total eclipse. What curious custom prevails among the Hindoos? What is the Saros? Is it now of any value? What is the metonic cycle? Explain its use. What is the golden number? What is the cause of a lunar eclipse? Why are lunar eclipses seen oftener than solar ones? How were total eclipses formerly regarded?

147. THE TIDES.†—Define ebb. Flow. How often does the tide happen? Explain the cause. Why does the tide occur about fifty minutes later each day? Why is there a tide on the side opposite the moon? The sun is much larger than the moon; why does it not produce the larger tide? What effect has the friction of the tides produced upon the earth? What theory upon this topic has Professor Ball advanced? What is meant by the differential effect of the moon? Why is not the tide felt out at sea? What is spring-tide? Neap-tide? Why does the tide differ so much in various localities? Tell about the height of the tides at different points. Why is there no tide on a lake? Is the tidal wave a forward movement of the water?

150. MARS.—Definition and sign? Describe the appearance of this planet. When is it brightest? What is its distance from the sun? Velocity? Day? Year? Distance from the earth? What is the peculiarity of its orbit? What is the diameter of Mars? Its volume and density as compared with the earth? How far would a stone fall on its surface the first second? Who discovered its moons? What is the peculiarity of these tiny globes? What are the peculiar telescopic features of Mars? What is the cause of its ruddy color? What are the snow-zones? Can we watch the change of its seasons?

* Lockyer, describing the beginning of a total eclipse, says: "One seems in a new world—a world filled with awful sights and strange forebodings, and in which stillness and sadness reign supreme; the voice of man and the cries of animals are hushed; the clouds are full of threatenings and put on unearthly hues; dusky, livid, or purple, or yellowish crimson tones chase each other over the sky irrespective of the clouds. The very sea is responsive and turns lurid red. All at once the moon's shadow comes sweeping over air, and earth, and sky, with frightful speed. Men look at each other and behold, as it were, corpses, and the sun's light is lost."—Gillis. in his observations upon the eclipse of 1859, as witnessed by him in Peru, remarks: "At 1.54, the moment of totality, the attendants, catching sight of the corona, dropped on their knees, and shouted, ' La Gloria ! La Gloria !'"

† As the tidal wave does not move as rapidly as the earth does, the water has an apparent backward motion. It has been suggested that this (as well as the friction of the atmosphere) acts as a break on the earth's diurnal revolution. It has been shown that the moon's true place can be best calculated if we suppose that the sidereal day is shortening at the rate of $\frac{1}{1000}$ of a second in 2,400 years. (See page 89.)

154. MINOR PLANETS (ASTEROIDS).*—Give Bode's law. Tell how the first of these planets was discovered. How many are now known? Are they probably all discovered? Describe some of these "pocket planets". Do they all lie within the Zodiac? What is their origin? (*Ans.* According to the nebular hypothesis, the ring of matter broke up into numberless small bodies, instead of aggregating into one large planet.) Give some of the names and signs.

157. JUPITER.—Definition and sign? Describe his appearance. Describe his orbit. What is his distance from the sun? Velocity? Day? Year? Distance from the earth? Diameter? Volume? Density? Centrifugal force? Force of gravity? Figure? Describe his seasons. Upon what does the change of seasons in any planet depend? What must be the appearance of the Jovian sky? Describe the telescopic features of Jupiter. Are Jupiter's moons visible to the naked eye? What are their names? What is their size? What space do they occupy? Describe the eclipse of Jupiter's moons. Define immersion, emersion, and transit. How rapidly do the satellites revolve? What can you say of the frequency of eclipses on Jupiter? Describe the belts. Why are they parallel to its equator? How was the velocity of light discovered? Does Jupiter emit light? Is it probable that a solid crust has formed over this planet? In what way is Jupiter reproducing the earth's history?

164. SATURN.—Definition and sign? Describe Saturn's appearance in the heavens. How rapidly does this planet move through the sky? What is its distance from the sun? What is the peculiarity of its orbit? What is its velocity? Year? Day? Distance from the earth? Diameter? Volume? Density? Force of gravity? Describe its seasons. Has it any atmosphere? Who discovered the rings of Saturn? Describe them. Which are the Bright Rings? Which is the Dusky Ring? Are they stationary? Explain their phases. Of what are they composed? Does Saturn emit light? Describe Saturn's belts. Describe Saturn's moons. The scenery on Saturn.

170. URANUS.—Definition and sign? How was this planet discovered? Tell of its previous observation by Le Monier. Is Uranus visible to the naked eye? What is its distance from the sun? Year?

* "It may surprise some persons to learn that the total mass of the two or three hundred small planets which have been discovered between the orbits of Mars and Jupiter, is sufficient only to make a globe a little over 400 miles in diameter. In other words if our globe were divided into 8,000 equal parts, one of these parts would equal in bulk and in weight the total of all these asteroids. Or, cut the earth through the equator, then take a section of about three-fourths of a mile in thickness, and it would furnish material for all these small planets and something remaining. It would seem that the solar system could not be much damaged, if some of these small planets should drop out of their courses and join some of the larger ones."

Diameter? Density? Describe its seasons. Telescopic features. Satellites. Peculiarity of its moons.

172. NEPTUNE.—Definition and sign? What is the appearance of this planet in the sky? Give an account of its wonderful discovery. What is its distance from the sun? Year? Velocity? Diameter? Volume? Density? Do we know anything of its seasons? Why not? What is the appearance of the heavens? What are the telescopic features of Neptune? Has Neptune any moon? What advantage have the Neptunian astronomers?

175. METEORS, AEROLITES, AND SHOOTING-STARS.—Define an aërolite. A shooting-star. A meteor. Give some account of the fall of aërolites. What elements are found in aërolites? How can an aërolite be distinguished? Give an account of wonderful meteors. Of shooting-stars.

176. Describe the showers of 1799 and 1833.* The shower of 1866. At what intervals did these showers occur? Why was not the shower of 1866 seen in this country? (*Ans.* Our side of the earth was not turned toward the meteors.) What is the average number of meteors and shooting stars daily? Why do we not see more of them? In what months are they most abundant?† Describe the origin of meteors and shooting-stars. What is their velocity? What causes the light? The explosion often heard? What is the theory of meteoric rings? What is their shape? How do these streams of meteoroids account for the showers at regular intervals? What is the period of the November ring? Why is the August shower so uniform, while the November one is periodic?‡ What is the relation between meteors and comets?

---

* A southern planter, describing the effect of the star-shower of 1833, says: "I was suddenly awakened by the most distressing cries that ever fell on my ears. Shrieks of horror and cries for mercy I could hear from most of the negroes of three plantations, amounting in all to about 600 or 800. While earnestly listening for the cause, I heard a faint voice near my door calling my name. I arose, and taking my sword, stood at the door. At this moment I heard the same voice still beseeching me to rise, and saying, 'Oh, my God, the world is on fire!' I then opened the door, and it is difficult to say which excited me most, the awfulness of the scene or the cries of the distressed negroes. Upwards of one hundred lay prostrate on the ground, some speechless, and some with the bitterest cries, with their hands raised, imploring God to save the world and them. The scene was truly awful, for never did rain fall much thicker than the meteors towards the earth: east, west, north, and south, it was the same."

† It has been noticed, from very early times, that the night of the 10th of August (St. Laurence's Day) is especially favorable for the occurrence of shooting-stars: and in Catholic Ireland, these stars, on the 10th of August, are always called the "tears of St. Laurence the Martyr," who was put to death by being broiled upon a gridiron.

‡ The fact that the November meteoroids are collected in a shoal instead of being distributed uniformly through the orbit gives color to the idea that this stream has not been long a member of the solar system. "In 1867, Leverrier stated his be-

What do you mean by the radiant point? What is the height of meteors? Weight?

185. COMETS.—How were comets looked upon by the ancients? Illustrate. Define the term comet. What is the tail?* The nucleus?

lief that the November meteors form the remains of some comet that had been recently introduced into the solar system by the attraction of one of the large outer planets. He found that the year A. D. 126 would give a position to the planet Uranus capable of producing such an effect, by converting the parabolic path of a comet into the path now described by the November meteors. In the year A. D. 137, the changed path of the comet for the first time came near the earth in her orbit round the Sun, since which year the petrified comet or shower of stones has completed 52 entire revolutions, the last of which terminated on the 13th of November, 1866. Theophanes of Byzantium relates that in November, A. D. 472, the sky at Constantinople appeared to be on fire with flying meteors. This corresponded with the tenth revolution of the November meteors.—Condé, in his history of the dominion of the Arabs, speaking of the year A. D. 902, states that in the month of October (13th), on the night of the death of King Ibrahim Ben Ahmed, an immense number of falling stars were seen to spread themselves over the face of the sky like rain, and that the year in question was thenceforth called the 'Year of Stars.' This year corresponded to the twenty-third revolution of the November meteors.—A similar shower of stars took place on the 17th of October, A. D. 934.—On the 14th of October. A. D. 1002, a remarkable shower of shooting-stars is noted by the Arab astronomers and historians, corresponding with the completion of the twenty-sixth revolution of the November meteors.—It is related in the annals of Cairo that on the 19th of October, A. D. 1202, the stars appeared like waves upon the sky, towards the east and west; they flew about like grasshoppers, and were dispersed from left to right. This shower corresponded with the thirty-second revolution of the November meteors.—On the 22nd of October, A. D. 1366, a shower of stars was noted, corresponding with the thirty-seventh revolution of the November meteors.—A similar phenomenon (forty-second revolution) was observed on the 25th of October, A. D. 1533.—The forty-seventh revolution was noted on the 9th of November, A. D. 1698.—The fiftieth revolution, observed by Humboldt and Boupland, on the 12th of November, A. D. 1799, as already remarked, first led modern astronomers to speculate on the true nature of these remarkable periodic phenomena.—The early observations of this meteoric shower were dated on the 12th of October, and during 52 revolutions the intersection of its orbit with that of the earth has moved on to the 14th of November.—Mr. Adams has shown this movement of nodes to be a consequence of the attractions of the superior planets, and has finally demonstrated the truth of the cometary origin of the November meteors."—*Houghton.*

* " Comets are almost always accompanied by tails, which are placed in the line joining the Sun and Comet, and on the side opposite to the Sun. Exceptions to this rule, though rare, sometimes occur. For example, the tail of the Comet of 1577 deviated 21° from the line joining the Sun and the Comet, and the tail of the Comet of 1680 diverged 5° from the same line. Comets have been occasionally observed with two tails, one in the usual position, and the other in nearly an opposite direction, or towards the Sun. The angle between the two tails, when such a phenomenon has been observed, has always been very considerable, varying from 140° to 170°. This rare phenomenon of two tails is supposed to be connected with certain rapid changes which the gaseous substance of the Comet is observed to undergo on approaching the Sun. There are many instances on record, in which the tails of Comets were observed to stretch through 100° of the celestial sphere, and the apparent

The head ? The coma? Does each comet necessarily possess all
these parts ? How would a mere round, fleecy mass be known to be
a comet ? What mistake did Herschel make in looking, as he supposed, at one of this kind (p. 171)? Where do comets appear ? In
what direction do they move ? How does a comet look when first
seen ? Describe the approach of a comet to the sun. Upon what does
the time of greatest brilliancy depend ? What do you say of the number of the comets ? What was Kepler's remark ? Why do we not see
them oftener? Where did Lockyer see one ? Describe the orbits of
comets. Which class has been calculated ? Which classes never return ? Describe the difficulty of calculating a comet's orbit. Name
the periods of some comets. What has been the distance from the sun
of some noted comets ? Velocity ? What do you say of the density of a
comet ? Illustrate. Is there any danger of our running against a
comet ? Do comets shine by their own or by reflected light ? Tell what
you can of their variation in form and dimensions. Give some account
of the comets of 1811, 1835, and 1843. For what is Biela's comet
noted ? (*Ans.* " A very remarkable phenomenon attended the perihelion
passage of this comet in the latter end of 1845. It became divided
into two comets, which did not again re-unite, but traveled along together in similar orbits. This unique phenomenon was noticed for
the first time in America on the 29th of December. The greatest distance observed between these two fragments of Biela's comet, before
their final disappearance, was about *two-thirds* of the moon's distance
from the earth.") For what is Encke's comet noted? What is its period ?
Give some description of Donati's comet. The comet of 1882.

196. ZODIACAL LIGHT.—Where can this be seen ? What is its appearance ? Where is it brightest? What is its origin ?

III. THE SIDEREAL SYSTEM.

203. Tell something of the appearance of the heavens at Neptune's
distance from the sun—our starting-point. Do we ever see the stars?
What do we see, then ? Which star is nearest the earth ? What is its
parallax? Its distance? How long would it take light to reach the
nearest star? How would the earth's orbit appear at that distance ? Our
sun? How long does it take for the light of the smaller stars to reach
the earth ? What can you say of the motion of the fixed stars? Illustrate.

length of the tail is known to undergo most rapid changes. We shall mention only
one case as an example of this phenomenon. The Comet of 1618 presented to the
Danish astronomer, Longomontanus, a tail of 104° in length, while it had been
measured by Kepler a few days previous, and ascertained to be only 70° long."

What proof have we that the fixed stars are suns ?* Describe the motion of the solar system. Is the center known? How many stars can we see with the naked eye? With a telescope? Have all the stars been discovered? What is the cause of the twinkling of the stars? Do we know anything of the magnitude of the stars? Name the points of difference between a planet and a fixed star. What do you mean by a star of the first magnitude? How many are there? Of the second magnitude? How many sizes can one see with the naked eye? What is the cause of the difference in the brightness? How are the stars named? Describe the division of the stars into constellations. Is there any real likeness to the mythological figures? Name any figure which seems to you well founded. Are the boundaries distinct? Who invented the system? State the possible meaning of the signs of the Zodiac and their origin. Explain why the signs and the constellations of the Zodiac do not agree. What causes the appearance of the constellations? Would they appear as they now do, if we should go out into space among them? Are the present forms permanent? State the value of the stars in practical life. What were the views of the ancients with regard to the stars? Describe the division of the stars into three zones.

214. THE CONSTELLATIONS.—The questions on these are uniform: (1) *description*, (2) *principal stars*, and (3) *mythological history*. Therefore, they need not be repeated with each constellation. What are the pointers? Does Polaris mark the exact position of the North Pole? How many times per day is Polaris on the meridian of any place? Explain how this applies in navigation or surveying. State how the amount of the variation from the true north will change through the ages. What star will ultimately become the pole-star? What curious facts are stated concerning the Pyramids? What do you say of the distance of Polaris? How may latitude be calculated by means of Polaris? What stars never set in our sky? What stars never rise?† Will the

* Sirius shines at least 200 times as brightly as our sun would shine if set beside it. Assuming its surface to be equally brilliant, this would imply, in comparison with our sun, a diameter 14 times and a volume 3,000 times as great. Its luster, however, seems higher than the sun's, but, even making allowance for that, we must still consider this giant sun to be at least 1,000 times as large as our own orb. Recent evidence tends to show that its rate of recession from us is diminishing, so that we may expect this to change into a motion of approach. Here is a hint that Sirius is travelling in a mighty orbit with movements carrying it alternately from and toward us.—*Proctor.*

† All stars whose north polar distance is less than the latitude of any place, will never set at that place, and all stars whose south polar distance is less than the latitude, will never rise. The Greeks and the Romans were familiar with the fact that certain stars never descend below the horizon. The following quotations are interesting:

" Immunemque æquoris Arcton."
                              OVID, METAM. xiii. 293.
        " Arctos
Æquoris expertes."   ID. 726—7.

Big Dipper always appear as now? Name three bright stars near the first meridian. (*Ans.* α Andromedæ, γ Pegasi, and β Cassiopeiæ.) How many degrees of longitude correspond to an hour of time? At what rate is Sirius receding from the earth? How has this motion been discovered? (See page 261.)

239. DOUBLE STARS, ETC.—Does any star appear double to the naked eye? How many have been found by the use of the telescope? What is an optical double star? Are all double stars of this class? Describe the revolution of a binary system. What other combinations have been discovered? What are their periods? Orbits? Mass? Are these companion stars as close to each other as they seem?

241. Name some prominent colored stars. Do their colors ever change? Which colors would indicate the hottest star? What is the probable effect in a system having colored suns?

242. What are variable stars? Describe the changes of Algol. Of Mira. What is the cause?

243. What are temporary stars? Describe the one seen in Cassiopeia. The one in Corona Borealis, in 1866? What are lost stars? Can you give any explanation of this phenomenon? Of what did the star of 1866 consist? Are these stars destroyed? Is the process of creation now complete?

245. What are star clusters? Name several. Is such a grouping a mere optical effect? Are they probably as closely compacted as they seem to be?

246. What are nebulæ? How do they differ from clusters? Is it probable that all nebulæ will be resolved into clusters? What is the general belief concerning nebulæ? What has spectrum analysis proved some of the nebulæ to be? Where are they most abundant? What can you say about their distances? Into how many classes, for convenience, are they divided? Describe and illustrate the elliptic nebulæ. What is said of the distance of the great nebula in Andromeda?

Ἄρκτον θ᾽ ἣν καὶ ἅμαξαν ἐπίκλησιν καλέουσιν,
Ἥτ᾽ αὐτοῦ στρέφεται, καί τ᾽ Ὠρίωνα δοκεύει,
Οἴη δ᾽ ἄμμορός ἐστι λοέτρων Ὠκεανοῖο.

ILIAD, xviii. 487—9. and ODYS. v. 273—5.

Ἄρκτοι κυανέου πεφυλαγμένοι Ὠκεανοῖο.

ARATUS, PHÆNOM. 48.

" Arctos oceani metuentes æquore tingui."

VIRG. GEORG. i. 246.

In order to understand the meaning of the expressions πεφυλαγμένοι Ὠκεανοῖο, and " *æquoris expertes*," as used by a Greek or Italian, we should remember that the north polar distance of η Ursæ Majoris is 39° 56′ 48″ ; and since the latitude of Athens is 37° 58′, and that of Naples 40° 50′, an inhabitant of the former city would see this star descend below the northern horizon for a small portion of its course ; and an inhabitant of Naples would see it sink within 3′ of the horizon, so as just to move along its northern edge.

The number of stars it contains? Describe the annular nebulæ. What is said of the "ring universe" in Lyra? Its diameter? Describe the spiral nebula in Canes Venatici. Describe the planetary nebulæ. What is said of the number and size of these "island universes"? Describe the fantastic appearance of the irregular nebulæ.

251. What are nebulous stars? What is their structure? What are variable nebulæ? Give instances. What is said of double nebulæ? Is anything definite known with regard to them? What are the Magellanic clouds?

253. Describe the Milky-way. What can you say of the number of stars in the Galaxy? Are the stars uniformly distributed?

254. What is Herschel's theory of the constitution of the universe? If this theory be true, what is our sun?

255. Give an account of the Nebular hypothesis. What is said of Saturn's rings? May they ultimately disappear?

259. What is spectrum analysis? Name the three kinds of spectra. What colored rays will a flame absorb?* Describe the spectroscope. What are Fraunhofer's lines? What is known of the constitution of the sun? What proof have we that iron exists in the sun? What elements have been found in the sun? What proof have we that the

* The power which gases possess of cutting out the particular lines which belong to the color that each emits has been beautifully illustrated by Prof. Newcomb. He says: "Suppose nature should loan us an immense collection of many millions of gold pieces, out of which we were to select those which would serve us for money, and return her the remainder. The English rummage through the pile, and pick out all the pieces which are of the proper weight for sovereigns and half-sovereigns; the French pick out those which will make five, ten, twenty, or fifty-franc pieces; the Americans the one, five, ten and twenty dollar pieces, and so on. After all the suitable pieces are thus selected, let the remaining mass be spread out on the ground according to the respective weights of the pieces, the smallest pieces being placed in a row, the next in weight in an adjoining row, and so on. We shall then find a number of rows missing: one which the French have taken out for five-franc pieces; close to it another which the Americans have taken for dollars; afterwards a row which have gone for half-sovereigns, and so on. By thus arranging the pieces, one would be able to tell what nations had culled over the pile, if he only knew of what weight each one made its coins. The gaps in the places where the sovereigns and half-sovereigns belonged would indicate the English, that in the dollars and eagles the Americans, and so on. If, now, we reflect how utterly hopeless it would appear, from the mere examination of the miscellaneous pile of pieces which had been left, to ascertain what people had been selecting coins from it, and how easy the problem would appear when once some genius should make the proposed arrangement of the pieces in rows, we shall see in what the fundamental idea of spectrum analysis consists. The formation of the spectrum is the separation and arrangement of the light which comes from an object on the same system by which we have supposed the gold pieces to be arranged. The gaps we see in the spectrum tell the tale of the atmosphere through which the light has passed us; in the case of the coins they would tell what nations had sorted over the pile."—*Newcomb's Astronomy*, p. 228.

stars are suns? What can you say of the similarity existing between the stars and our earth? What has been discovered with regard to the constitution of the Nebulæ? Of their relative brightness? How has the proper motion of the stars been shown?

263. TIME.—What two methods of measuring time? What is a sidereal day? What are astronomical clocks? Tell how they are used. Why do astronomers use sidereal time? What is a solar day? What causes the difference between a sidereal and a solar day? To how much time is a degree of space equal? Which is taken as the unit, the solar or the sidereal day? How long is a solar day? A sidereal day? A solar day equals how many sidereal hours? A sidereal day equals how many solar hours? Describe mean solar time. What is apparent noon? Mean noon? The equation of time? When is this greatest? When least? When do mean and apparent time coincide? Can a watch keep apparent time? How may apparent time be kept? How can it be changed into mean time? Tell how to erect a sun-dial. When will a sidereal and a mean time clock coincide? A mean-time clock and the sun-dial? How did the ancients measure time, before the invention of clocks and watches?* State the two reasons why the solar days are of unequal length. What is the civil day? Who invented the present division? Describe the customs of various nations. What

---

* " The ancients used clepsydræ and sun-dials, to measure time. The clepsydræ, in its simplest form, resembled the hour-glass, water being used instead of sand, and the flow of time being measured by the flow of the water. After the era of Archimedes, clepsydræ of the most elaborate construction were common; but while they were in use, the days, both winter and summer, were divided into twelve hours from sunrise to sunset, and consequently the hours in winter were shorter than the hours in summer; the clepsydra, therefore, was almost useless except for measuring intervals of time, unless different ones were employed at different seasons of the year. The sun-dial was a great improvement upon the clepsydræ; but at night and in cloudy weather it could not be used, of course, and the rising, culmination, and setting of the various constellations were the only means available for roughly telling the time during the night. Indeed, Euripides, who lived 480-407 B. C., makes the Chorus in one of his tragedies ask the time in this form :—

' What is the star now passing ?'

and the answer is :—

' The Pleiades show themselves in the east ;
The Eagle soars in the summit of heaven.'

It is also on record that as late as A. D. 1108 the sacristan of the Abbey of Cluny consulted the stars when he wished to know if the time had arrived to summon the monks to their midnight prayers; and in other cases, a monk remained awake, and to measure the lapse of time repeated certain psalms, experience having taught him in the day, by the aid of the sun-dial, how many psalms could be said in an hour. When the proper number of psalms had been said, the monks were awakened."— *Lockyer.*

is the origin of the names of the days?* What is the sidereal year? The mean solar year? What causes the difference? What is the anomalistic year? How did the ancients find the length of the year? What error did they make? What was the result? Give an account of the Julian calendar. The Gregorian calendar. What is the meaning of the terms O. S. and N. S.?† What country now uses O. S.? When was the change adopted in England? ‡ How was it received? How could a child be eight years old before a return of its birthday? When do the Jews begin their year? Why does our year begin January 1st? Show how the earth is our timepiece. What influence has Jupiter's moons on the cotton trade?

CELESTIAL MEASUREMENTS.—These problems are to be used throughout the study. They require no questions but the formal statement of the problem requiring solution.

* It is said that the Egyptians named the seven days from the seven celestial bodies then known. The order was continued by the Romans Tuesday they called *Dies Martis;* Wednesday, *Dies Mercurii;* Thursday, *Dies Jovis;* Friday, *Dies Veneris.* In the Saxon mythology. Tius. Woden, Thor, and Friga are equivalent to Mars, Mercury. Jupiter, and Venus. Hence we see the origin of our English names.

† " As an illustration of the effect of the change of *style,* we may instance the case of Washington. He was born February 11, 1732, before the change of style. Inasmuch as 1752 began on the 25th of March and ended on the 31st of December, he had no birth-day in that year ; hence. he was 20 years old on the 22nd of February, 1753, new style. Because anniversaries are always determined according to the civil calendar, the birth-day of Washington is properly celebrated on the 22nd of February, and not on the 23d, as some have contended, on account of the day dropped in the year 1800."—*Peck's Astronomy,* p. 216.

‡ " In England. from the 14th century till the change of style in 1752. the legal and the ecclesiastical year began March 25. After the change was adopted in 1752, events which had occurred in January, February, and before March of the old legal year, would. according to the new arrangement, be reckoned in the next subsequent year. Thus the revolution of 1688 occurred in February of that legal year, or, as we should now say, in February, 1689 ; and it was, at one time, customary to write the date thus: February, 168⅞."—Appleton's Cyclopædia, article on Calendar.

GUIDE TO THE CONSTELLATIONS.

THE following is a description of the appearance of the heavens on or about the first day of each month in the year.

January. (7 P. M.)—*In the North*, Cassiopeia and Perseus are above Polaris, Cepheus and Draco west, Ursa Minor is below, and Ursa Major below and to the east. *In the East*, Cancer is just rising, Canis Minor (Procyon) has just risen. *Along the Ecliptic*, Gemini is well up, then Taurus, Aries—reaching to the meridian, next Pisces; Aquarius (letter Y) and Capricornus are just setting. *In the Southeast*, Orion and the Hare are well up. *In the South*, Cetus swims his huge bulk far to the east and west. *In the Southwest*, is Piscis Australis (Fomalhaut). *North of the Ecliptic*, the Triangles are nearly in the zenith, Perseus is just east, below is Auriga, Andromeda lies just west of the meridian, and Pegasus is midway; Delphinus (the Dolphin, Job's Coffin), Aquila (Altair), and Lyra (Vega) are fast sinking to the western horizon; while, along the Milky Way, blazes the brilliant cross of Cygnus.

February. (7 P. M.)—*In the North*, Ursa Major lies east of Polaris, Ursa Minor and Draco are below, Cepheus is west, Cassiopeia above and to the west. *In the East*, Regulus and Cor Hydræ are just rising. *Along the Ecliptic*, Leo (Regulus, the sickle) just rising, Cancer well up, Gemini midway, Taurus on the meridian, Aries (the scalene triangle) past, Pisces next, and, lastly, Aquarius just setting. *In the Southeast*, Canis Minor, Canis Major (Sirius), and Orion are conspicuous. *In the Southwest*, Cetus covers nearly the whole sky. *North of the Ecliptic*, Perseus is on the meridian, while Auriga is a little east of it; west of Perseus is Andromeda, while the Great Square of Pegasus is fast approaching the horizon. *In the Northwest*, Cygnus is setting.

March. (7 P. M.)—*In the North*, Ursa Major lies east of Polaris, Draco and Ursa Minor are below, Cepheus is below and to the west, and Cassiopeia west. *In the East*, Cor Caroli is well up, toward the northeast, and Coma Berenices is rising. *Along the Ecliptic*, Leo is fully risen, next Cancer, Gemini reaches to the meridian, Taurus is past, Aries midway, and, lastly, Pisces is just beginning to set. *In the Southeast*, Cor Hydræ, Canis Minor, and Canis Major are conspicuous. *In the South*,

Orion blazes brilliantly. *In the Southwest,* Cetus is hiding below the horizon. *North of the Ecliptic,* Auriga is in the zenith; west are Perseus and Andromeda, while Pegasus is just beginning to sink out of sight.

April. (7 P. M.)—*In the North,* Ursa Major is above and to the east of Polaris ; opposite and to the west is Perseus, Draco below and to the east, Cepheus below and to the west, Cassiopeia west. *In the East,* Boötes (Arcturus) is not quite fully r.sen. *Along the Ecliptic,* Virgo (Spica) is rising, Leo midway, Cancer reaches to the meridian, Gemini is past, next Taurus, then Aries, and, lastly, Pisces just setting. *In the Southeast,* is the Crater (the Cup) ; Hydra stretches its long neck to the meridian. *In the South,* Canis Minor. *In the Southwest,* Sirius and Orion ; the Egyptian X (p. 229) can now be seen. *North of the Ecliptic,* and in the northeast, are Coma Berenices and Cor Caroli ; above Gemini and Taurus is Auriga, while Andromeda is just setting in the northwest.

May. (8 P. M.)—*In the North,* Ursa Major is above Polaris, Ursa Minor and Draco are east, Cepheus and Cassiopeia below, and Perseus is west. *In the East,* Lyra is rising, and Hercules is just up. *Along the Ecliptic,* Libra is just rising, Virgo is midway, Leo is on the meridian, Cancer is past, next Gemini, and lastly Taurus just setting. *In the South,* stretching east and west of the meridian, is Hydra, with the Crater and Corvus a little east. *In the Southwest,* are Cor Hydræ, Canis Major, and Canis Minor, while Orion is just setting in the west. *North of the Ecliptic,* in the east, above Hercules, are Corona Borealis (The Northern Crown), Boötes (Arcturus), Coma Berenices, and Cor Caroli, which stretch nearly to the meridian. *In the Northwest,* above Taurus and Perseus, is Auriga.

June. (8 P. M.)—*In the North,* Ursa Major is above Polaris, Draco and Ursa Minor are east, Cepheus is below and east, and Cassiopeia directly below. *In the East,* Cygnus (the Cross) and Aquila are rising, Lyra and Taurus Poniatowskii are well up. *Along the Ecliptic,* Scorpio is rising, Libra is midway, Virgo on the meridian, Leo past, Cancer midway, Gemini next, and Taurus just setting. *In the South,* are Corvus and the Crater a little past the meridian. *In the Southwest,* is Cor Hydræ, and in the west Canis Minor is nearing the horizon. *North of the Ecliptic,* in the east, above Scorpio, is Hercules ; then Corona and Boötes, and, near the meridian, Cor Caroli. and Coma Berenices. *In the Northwest,* is Auriga, just coming to the horizon.

July. (9 P. M.)—*In the North,* Draco and Ursa Minor are above Polaris, Ursa Major is west, Cepheus east, and Cassiopeia below to the east. *In the East,* the Dolphin (Job's Coffin) is now well up, Cygnus is almost midway to the meridian, and Lyra is still higher. *Along the Ecliptic,* Capricornus is rising, Sagittarius (the Archer) is next, Scorpio, with its long tail swinging along the horizon, is directly south, Libra is past the meridian, Virgo midway, and Leo has almost reached the horizon. *In the Southwest,* the Crater is setting, and Corvus is just above. *North of the Ecliptic,* above Scorpio and east of the meridian, are Serpentarius, Hercules, and Taurus Poniatowskii ; Corona is almost on the meridian, to the west of which lie Boötes, Cor Caroli, and Coma Berenices.

August. (9 P. M.)—*In the North,* Draco and Ursa Minor are above Polaris, Cepheus is above and to the east, Cassiopeia east, and Ursa Major west. *In the Northeast,* Perseus is just rising, while south of it, Andromeda and Pegasus are fairly up. *Along the Ecliptic,* Aquarius is risen, next Capricornus, Sagittarius reaches to the meridian. Scorpio is just past, Libra next, and Virgo (Spica) just touches the horizon. *North of the Ecliptic,* Taurus Poniatowskii is on and Lyra is just east of the meridian ; the Swan and Dolphin are east of Lyra, Serpentarius and Hercules are above Scorpio, and just west of the meridian ; thence west are Corona and Boötes, while far in the northwest are Coma Berenices and Cor Caroli.

September. (8 P. M.)—Draco is above and to the west of Polaris, Cepheus above and to the east, Cassiopeia east, Ursa Major is below and to the west. *In the Northeast,* Perseus is just rising. *In the East,* Andromeda is fairly up, Pegasus is nearly midway to the meridian. *Along the Ecliptic,* Pisces is just rising, next Aquarius, Capricornus in the southwest, Sagittarius on the meridian in the south, next Scorpio in the southwest, Libra, and, lastly, Virgo just setting. *North of the Ecliptic,* Lyra is on the meridian, Cygnus, the Dolphin, and Aquila are just to the east ; while to the west, are Taurus Poniatowskii and Serpentarius ; north of these latter are Hercules, Corona, Boötes, Cor Caroli, and Coma Berenices.

October. (7 P. M.)—*In the North,* Cepheus and Draco are above Polaris, Ursa Minor is west, Cassiopeia east, and Ursa Major below and west. *In the Northeast,* Perseus is fairly risen. *In the East,* Andromeda is nearly midway to the zenith. *Along the Ecliptic,* Aries is just rising, Pisces well up, Aquarius and Capricornus are in the southeast, Sagittarius is in the south, Scorpio far down in the southwest, and Libra just setting. *North of the Ecliptic,* Cygnus and Aquila are on the meridian ; the Dolphin is just east of it, and far south ; Lyra is west of the meridian ; Taurus Poniatowskii is lower down and to the south ; Serpentarius is just above Scorpio ; next, in line with Scorpio and Polaris, is Hercules ; Corona and Boötes are toward the northwest, where Coma Berenices is just setting.

November. (7 P. M.)—*In the North,* Ursa Major is below Polaris, Ursa Minor and Draco are to the west, Cepheus is above, and Cassiopeia above and to the east. *In the Northeast,* Auriga is just rising, and Perseus is above, nearly midway to the meridian. *Along the Ecliptic,* Taurus is just rising, next are Aries and Pisces ; Aquarius is on the meridian, south ; then Capricornus, and lastly Sagittarius, in the southwest. *North of the Ecliptic,* Pegasus and Andromeda lie east of the meridian, the Swan, Dolphin, Eagle, Taurus Poniatowskii, and Lyra west. *In the Northwest,* are Hercules and Corona.

December. (7 P. M.)—*In the North,* Cassiopeia is above Polaris, Cepheus above and to the west, Perseus above and to the east, Draco west, and Ursa Major below. *In the Northeast,* below Perseus, is Auriga. *In the East,* Orion is rising. *Along the Ecliptic,* Gemini is just rising. Taurus is nearly midway, next Aries, Pisces is on the meridian, then Aquarius, and lastly Capricornus, far in the southwest. *In the South,* east of the meridian, is Cetus, and west is Fomalhaut. *North of the Ecliptic,* Andromeda is nearly on the meridian, and Pegasus west of it ; Cygnus, Delphinus, Lyra, and Aquila are about midway, while Taurus Poniatowskii is just sinking to the horizon. *In the Northwest,* Hercules is just setting.

NOTE.—It should be borne in mind that a month makes a variation of about two hours (30°) in the rise of a star ; hence, in the foregoing " Guide," the "January Sky" of 9 P. M. would be about the same as the " February Sky " of 7 P. M. ; the "January Sky" of 11 P. M. would be about the same as the " March Sky " of 7 P. M., &c. In this way the "Guide" may be used for any hour in the night. The pupil will see that in the "Guide" the prominent figures and stars in each constellation are given in parentheses. Examples: the " Y " in Aquarius, the " scalene triangle " in Aries, "Job's Coffin " in the Dolphin, " Procyon " in Canis Minor, &c. These aid in identifying the constellation.

APPARATUS.

To ILLUSTRATE PRECESSION OF THE EQUINOXES,* make the simple apparatus shown in the cut. It represents Fig. 38, and the explanation of that figure and several subsequent ones applies to it. The ingenuity of pupil and teacher will devise methods of explaining by means of this instrument many otherwise abstruse points under this difficult subject. The following are suggestive merely :

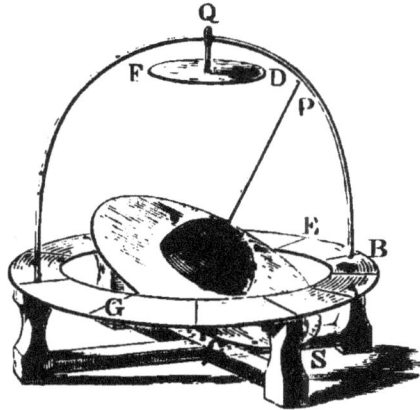

Fig. 114

1. *To show motion of earth's axis around pole of Ecliptic.*—Move P, axis of earth's plate, around D F, whose circumference roughly represents the little circle (ellipse) described by the pole of the earth. (Fig. 41.)

2. *To show change of Polar Star.*—The pupil can readily see that the north pole of the earth will, at different times, point to different stars located around this circle. Now, Polaris; next, Lyra.

3. *To show why present polar distance will gradually diminish and then increase* (p. 217).—The polar star lies at a little distance from this circle (edge of plate) and the pole is gradually approaching the star, but will pass it and then recede further from it, until, finally, Lyra, lying 5˚ from this circle, will become the polar constellation.

4. *To show Precession of Equinoxes.*—Pass axis of earth around small

---

* The above apparatus was devised by Solomon Sias, A.M. M.D., Principal of Schoharie Union School, N. Y. It can be made by any ingenious pupil. The plates are cut out of tin ; the standard may be made with the knife or scroll-saw to suit one's taste ; the earth is half of a little wooden ball balanced on the wire pin C ; and the semicircle, poles, etc., are of wire. The different parts may be soldered or fastened together with tacks.

circle. Note the position of the equinoxes before moving, and their gradual change of position along the ecliptic.

5. *To show cause of Precession.*—Apply explanation of **Fig.** 39.

6. *To show necessity for new stellar maps occasionally,* or careful reductions to previous standards. With the change of equinoxes, there is also a change of the equinoctial system, p. 27.

7. *To illustrate Fig.* 40.—Remove the wire semicircle, and, inclining the axis of the earth, spin the wire between the thumb and finger like a top. The equinoxes will pass around the ecliptic as they did when the axis was carried around in the previous experiments.

Letting G B E represent the plane of the ecliptic, and G E the line of the equinoxes, we can use this apparatus to illustrate the seasons, etc., (p 95). Also, by placing a lamp near S, the phenomenon of day and night, long summer days, short winter days, etc. (pp. 97, etc.), can be easily explained.

LIST OF INTERESTING OBJECTS VISIBLE WITH ORDINARY TELESCOPE.

COMPILED BY LEWIS SWIFT, F.R.A.S., DIRECTOR OF WARNER OBSERVATORY, ROCHESTER, N. Y.

Constellation.	Object.	R. A.		Dec.		Magnitudes.	Distance.	Remarks.
		h.	m.	°	'		"	
Andromeda—Double *	γ	1	57	+41	45	3.5 5.5	11	Triple. Beautiful object. Different colors.
" Nebula "	Messier 31	0	36	+40	35	- -	-	One of the grandest known. Visible to naked eye.
" "	M. 32	-	-	-		- -	-	In same field with above; small, bright.
Aquarius—D. *	ζ	22	23	—0	38	4 4.5	3	Binary. Fine object.
" Nebula	M. 2	21	27	—1	22	- -	-	Large. Stars of 15th magnitude.
Argo Navis—Cluster	- -	7	31	—14	12	- -	-	Fine cluster. Visible to naked eye.
" Nebula	M. 46	7	36	—14	32	- -	-	Stars of 10th magnitude.
Aries—D. *	γ	1	47	+18	42	4.5 5	8	First double star discovered. Fine object.
Boötes—D. *	ε	14	40	+27	35	3 7	3	Fine pair. Difficult. Companion, blue.
" D. *	ξ	14	46	+19	36	3.5 6.5	4	Fine pair. Binary. Period about 150 years. (?)
Cancer—D. *	ζ	8	5	+18	1	6, 7, 7.5	5, 1.3	Binary. D. * only with small telescopes.
" Cluster	M. 44	8	33	+20	22	- -	-	Easily resolvable. Fine object.
Canes Venatici—D. *	Cor Caroli.	12	50	+38	58	2.5 6.5	20	Fine pair. Easy.
" Cluster	M. 3	13	37	+29	56	- -	-	Resolvable into 1000 stars by large telescopes.
" Nebula	M. 63	13	25	+47	49	- -	-	Double. Lord Rosse's wonderful spiral.
" "	U.	13	11	+42	40	- -	-	Bright; very elongated; many nebulae near.
Cephens—D. *	β	21	27	+70	2	3 8	13	
" D. *	δ	22	25	+57	48	4.5 7	41	Faint star in center of a faint, large, round nebula.
" Nebulous *	- -	21	2	+67	38	- -	-	
" Variable *	U.	0	52	+81	14	- -	-	From 7 to 9 in 2d 11h 50m. The 2d most rapid known.

LIST OF INTERESTING OBJECTS, ETC.—Continued.

Constellation.	Object.	R. A. h.	R. A. m.	Dec. °	Dec. '	Magnitudes.		Distance.	Remarks.
Cetus—Variable *	Mira	2	11	— 3	31	—	—	"	From 2d to 10th magnitude in 331 days.
" D. *	R.	2	30	— 0	43	—	—	—	From 8th to 13th magnitude in 167 days.
" Nebula		2	42	— 25	11	—	—	—	Very bright; very elongated; very large.
Comæ Berenices—D. *		12	29	+19	2	5.5	7	20	
" D. *		12	15	— 27	42	7	7	8.5	
" Variable *	R.	12	56	+19	57	—	—	—	From 7th to 13th magnitude in 363 days.
" Nebula	M. 53	13	7	+18	47	—	—	—	Mass of stars of 11th to 15th magnitude.
" "	M. 64	12	51	+22	30	—	—	—	Large, bright. Many in this constellation.
Cygnus—D. *	β	19	26	+27	42	3	7	34	Yellow and blue. Very fine pair.
" "	61	21	1	+38	9	5.5	6	19	Second nearest star to us.
" "		20	4	+35	7	7	9	6.5	
" Variable *	R.	19	34	+57	10	8	8	4.5	From 6th to 13th magnitude in 425 days.
" "	S.	20	3	+57	38	—	—	—	From 9th to 13th magnitude in 322 days.
Delphinus—D. *	γ	20	41	+15	42	4	7	12	Beautiful pair.
" "		20	26	+10	51	7	8	23	
" "		20	11	+15	27	7	8	4	
" Variable *	R.	20	9	+8	41	—	—	—	From 8th to 13th magnitude in 284 days.
" "	T.	20	40	+15	58	—	—	—	From 8th to 13th magnitude in 203 days.
" Nebula		20	28	+6	6	—	—	—	Bright. Resolvable.
Draco—D. *	μ	17	3	+54	38	4	4.5	2.5	Binary. Period, 600 years. (?) Interesting pair.
" "		18	9	+79	59	5.5	6	20	Very bright. Resolvable.
" Variable *		14	37	+58	32	7	8	7.5	
" Nebula		15	3	+56	14	—	—	—	
Gemini—D. *	Castor	7	27	+32	9	3	3.5	4.5	Finest double star known. Easy.
" "		7	26	+23	7	6.5	8.5	11	
" "		7	4	+22	29	7.5	7.5	8.5	Fine pair.

Constellation.	Object.	R. A. h. m.	Dec. ° '	Magnitudes.	Distance.	Remarks.
Gemini—Variable *		7 0	+22 53			From 7th to 12th magnitude in 371 days.
" " Variable *		7 42	+24 2			From 8th to 13th magnitude in 298 days.
" Nebula	M. 35	6 1	+24 21			Cluster of minute stars.
Hercules—D.*	α	17 9	+14 32	3.5	4.5	Very interesting pair.
" "	ρ	17 20	+37 15	4 5.5	3.5	Fine pair.
" "		17 56	+21 36	5.5 5.5	6	
" "		18 3	+26 5	7 7	14	
" Variable *		16 5	+18 43			From 8th to 13th magnitude in 319 days.
" "		18 1	+31 0			From 7th to 12th magnitude in 165 days.
" Cluster	M. 13	16 37	+36 41			Many thousand stars. Wonderful object.
" "	M. 92	17 13	+43 16			Many thousand stars.
Hydra—D.*	ε	8 40	+ 6 51	4 8	3.5	Binary. Period between 500 and 600 years.
" "		8 29	+ 7 2	6 7	10.5	Fine pair.
" "		8 39	- 2 10	7 8	5	Fine pair.
" "		13 23	-22 40			From 4th to 10th magnitude in 469 days.
" Variable *		8 47	+ 3 31			From 7th to 12th magnitude in 256 days.
Leo—D.*	γ	10 13	+20 27	2 4	3.3	Binary. Interesting pair.
" "		10 49	+25 23	4.5 7	6	Binary.
" Variable *		9 41	+11 59			From 5th to 10th magnitude in 312 days.
" Nebula *	M. 65,66	11 14	+13 39			Two in one field.
Lyra—Double Double *	1 and 2	16 40	+39 33	5 5.5	{3.3 / 2.5}	Binaries. Finest object of the kind in the heavens.
" Nebula	M. 57	18 32	+41 13	7.5 7.5	6	Annular. Remarkable object.
" "		18 40	+32 53			
Monoceros—Triple *		6 23	- 6 57	6, 7, 8	7, 10	Most beautiful trio visible from this latitude.
Opbiuchus—D.*		17 8	-26 25	5 6	5	Binary. Period, about 80 years.
" "		17 59	+ 1 32	4.5 6	3	Resolvable cluster, 10th to 16th magnitude.
" Nebula	M. 12	16 41	+ 2 45			Resolvable cluster, 11th to 15th magnitude.
" "	M. 10	16 51	- 3 56			From 6 to 7 in 2h 7m 41s. The shortest period variable [known.
" Variable *		17 10.5	+ 1 20.5			

LIST OF INTERESTING OBJECTS, ETC.—Concluded.

Constellation	Object	R.A. (h. m.)	Dec. (° ')	Magnitudes	Distance (″)	Remarks
Orion—Triple *	ζ	5 35	— 2 00	3, 6.5, 10	2.5, 56	Beautiful object.
"	ι	5 30	— 5 59	3.5, 8.5, 11	11, 49	Beautiful object.
" D. *	σ	5 33	— 2 40	4, 8, 7	12, 42	
" Nebula	M. 42	5 29	+ 1 24	6, 7	21	Greatest marvel in the heavens.
Perseus—Variable *	Algol.	3 0	+40 30			From 2d to 4th magnitude. Most remarkable variable known.
Sagittarius—Nebula	M. 22	18 29	—24 0			Cluster of stars of 11th magnitude.
"	M. 25	18 25	—19 10			Coarse cluster.
"	M. 8	17 57	—24 21			Visible to naked eye.
Scorpio—D. *	β	15 58	—19 29	2, 5.5	13	
" Triple *	ξ	15 58	—11 2	4.5, 5, 7.5	1, 7	
" Variable *	T.	16 10	—22 41			From 7th to 10th magnitude, irregular.
" Nebula	M. 80	16 10	—22 42			The above variable is in the middle of the nebula.
Ursa Major—D. *	ζ	11 19	+55 33	3	14	Interesting object for small telescopes.
" Variable *	—	11 32	+45 46	6	10	From 6th to 12th magnitude in 255 days.
" Nebula	M. 81	10 36	+69 21			Bright. Resembles a comet. M. 82 near.
		9 46	+69 41			
Virgo—D. *	γ	12 36	— 0 47	4	5	One of the most interesting pairs known. Binary.
" Variable *	R.	12 32	+ 7 39			From 6th to 11th magnitude in 145 days.
"	S.	13 27	— 6 35			From 5.5th to 18th magnitude in 374 days.
" Nebula	M. 86	12 20	+13 36			Resolvable ; one of 7 in one field.
Vulpecula—Nebula	M. 27	19 54	+22 23			The celebrated Dumb-bell nebula.

INDEX.

www.ingramcontent.com/pod-product-compliance
Lightning Source LLC
Chambersburg PA
CBHW021451210326
41599CB00012B/1027